MW00997421

Advance Praise for *Six Sigma for Sustainability*

"This book illustrates how quality and sustainability are converging as an unstoppable force for innovation, process improvement, and organizational change."

WILLIAM MCDONOUGH, DESIGNER, ARCHITECT,
CO-AUTHOR OF *CRADLE TO CRADLE*

"This is an incredibly useful guide to getting a substantive sustainability program going in a company—compelling writing, great examples and a solid environmental foundation. If you want to know what to do, how to do it, and can leverage your knowledge of Six Sigma, there's no better place to begin than here."

ELIZABETH STURCKEN
MANAGING DIRECTOR, CORPORATE PARTNERSHIPS
ENVIRONMENTAL DEFENSE FUND

"Jordan, McCarty, and Probst tackle the corporate sustainability challenge with anchors in shareholder value, customer requirements, measurements, and change management. For any leader that wants to run his or her business processes more sustainably, this is a critical read."

JUDAH SCHILLER, CEO
SAATCHI & SAATCHI S

"All MBAs would benefit in reading this book. It's important knowledge for helping future business leaders integrate sustainability into their toolkit."

LIZ MAW, EXECUTIVE DIRECTOR
NETIMPACT

"As every sustainability leader at a large company knows, delivering great results at scale requires excellent process management. This book shows how to achieve scale and measurable results to address one of the greatest challenges of our time."

LISA SHPRITZ, USGBC BOARD OF
DIRECTORS MEMBER, 2006–2011

About the Authors

Thomas McCarty, Six Sigma Master Black Belt, is a Managing Director within the Strategic Consulting group of Jones Lang LaSalle where he leads Six Sigma deployment strategies within the firm and for client engagements. He is the lead author of *The Six Sigma Black Belt Handbook*, and co-author of *Six Sigma Financial Tracking and Reporting*. Mr. McCarty's clients have included HSBC, United Health Group, Fidelity, Sprint, Agilent, American Electric Power, Starbucks, Motorola, and the U.S. Department of Defense.

Michael Jordan, Six Sigma Master Black Belt, is a senior vice president at Jones Lang LaSalle and leads their Americas corporate sustainability consulting practice. He has worked with executives at companies such as Cisco, The Coca-Cola Company, and Deutsche Bank to cut costs, manage carbon, and build greener buildings. Mr. Jordan is a LEED Accredited Professional, a trained facilitator in The Natural Step sustainability framework, and holds an MBA from the University of Colorado.

Daniel Probst has been in the commercial real estate industry for more than 30 years, including 23 with Jones Lang LaSalle. He currently serves as chairman of the global energy and sustainability services practice. Mr. Probst is responsible for developing and delivering products and services to help clients reduce energy costs and their real estate-related environmental footprint through innovative portfolio and occupancy strategies, workplace standards, and operating practices. He has a mechanical engineering degree from Purdue University and an MBA from Indiana University, and is a LEED Accredited Professional.

Six Sigma for Sustainability

How Organizations Design and Deploy Winning Environmental Programs

Thomas McCarty
Michael Jordan
Daniel Probst

New York Chicago San Francisco
Lisbon London Madrid Mexico City
Milan New Delhi San Juan
Seoul Singapore Sydney Toronto

The McGraw-Hill Companies

Cataloging-in-Publication Data is on file with the Library of Congress.

McGraw-Hill books are available at special quantity discounts to use as premiums and sales promotions, or for use in corporate training programs. To contact a representative please e-mail us at bulksales@mcgraw-hill.com.

Six Sigma for Sustainability

1 2 3 4 5 6 7 8 9 0 QFR/QFR 1 9 8 7 6 5 4 3 2 1

ISBN 978-0-07-175244-2
MHID 0-07-175244-7

The pages within this book were printed on acid-free paper containing 100% postconsumer fiber.

Sponsoring Editor
Judy Bass

Editorial Supervisor
David E. Fogarty

Project Manager
Patricia Wallenburg

Copy Editor
Jim Madru

Proofreader
Claire Splan

Indexer
Claire Splan

Production Supervisor
Richard C. Ruzycka

Composition
TypeWriting

Art Director, Cover
Jeff Weeks

To our children: Shannon, Michael, Nathan, Dan,
Kyle, Laura, Mick, Cooper, Reed, and Zack.

CONTENTS

PROLOGUE

TOM: I am a slow runner, a short-distance cyclist, and a high-handicap golfer. I pursue all three sports with vigor and frustration. My pursuits are based on three basic needs: the need to stay fit, the need to compete, and primarily, the need to be outside. I need fresh air and sunshine to stay alive. It turns out, we all do. However, as much as I enjoy and thrive on fresh air and sunshine, it is only in the past 10 years that I have come to realize that the environment, which I enjoy and much depend on, is fragile and in need of dramatic repair.

In fact, it wasn't so long ago that my daughter was scolding me because I wouldn't even take the time to separate my beer cans from other trash. And she would stand behind me and turn the water off when I was brushing my teeth. I just wasn't taking personal responsibility for my contribution to building a better environment. Much of my apathy was due to the fact that I really didn't believe I could make a difference.

So my journey toward being a sustainability change agent has taken an interesting path. I spent 28 years of my career at Motorola. In the late 1980s and 1990s, my colleagues at Motorola and I learned that very complex business problems, which at the time seemed insolvable, could be solved by using a structured problem-solving approach supported by powerful new statistical analysis tools and enlightened leadership. That problem-solving approach evolved into the methodology that we now call *Six Sigma.* I witnessed some amazing changes in the way we were doing business at Motorola, brought on through statistical analysis and fact-based decision making. A team transformed the delivery cycle time of small communications devices from 8 weeks to 48 hours and dropped the product-development cycle time of certain mobile phones from 3 years to 18 months. Another team used the Six Sigma methodology to transform a toxic solder bath into an orange juice–based, environmentally friendly bath. During the early 2000s, I became director of Six Sigma for customers and suppliers at Motorola University. In that role, I had the opportunity to spread the Six Sigma gospel to Motorola's key customers and suppliers. We taught suppliers how to reduce the variation in their products and improve their cycle times. We also used the methodology to improve their scrap sorting processes and convert electronic waste (which, at the time, was going to landfill) into cash. For one supplier, we generated over $1 million per year in cash through improved recycling. I learned, in that effort, that it wasn't just about doing things that were good for the environment. It had to be good for the business.

That work at Motorola University led me to Jones Lang LaSalle, a global provider of real estate services. At the time, Jones Lang LaSalle was Motorola's real estate outsource partner, managing Motorola's facilities and construction processes worldwide. I helped Jones Lang LaSalle's senior leadership team understand how the Six Sigma problem-solving methodology and management frameworks could be applied to the service-delivery processes associated with acquiring, building, and managing corporate real estate portfolios. Along the way, I became managing director of Jones Lang LaSalle Strategic Consulting and the Six Sigma practice leader for the firm.

In an early Six Sigma project at Jones Lang LaSalle, I worked with a team at a large financial institution to better understand how to reduce energy costs across a footprint of more than 1,000 branch banks. The Six Sigma methodology helped the team understand how to measure and then prioritize their opportunities to reduce energy consumption. More important, Six Sigma helped the team build a compelling business case for installing monitoring and control systems and eliminating inefficient heating and cooling systems. While senior leaders were interested in improving their carbon footprint, it was the money that got them focused and motivated. The changes saved them $5 million in the first year. More recently, I have been involved in a multiyear effort to reduce the carbon footprint created by the construction and management of data centers. Using small teams within each data center and employing a common set of Six Sigma measurement and analysis tools, we have found ways to reduce both water usage and electricity consumption. Early results demonstrate an annual savings of $100,000 per data center.

So I have come to learn that by using Six Sigma methodology, I can have a substantial impact on the environment. It may not improve my golf game or make me a faster runner, but now I can look my daughter in the eyes and feel proud to be doing my share to drive sustainability across a significant commercial real estate portfolio.

My contributions to this book are my attempt to share my experience with using the Six Sigma methodology to drive wide-scale corporate sustainability campaigns. I have learned that success will come through a collaborative engagement of leadership, improved change management, proactive use of focused teams, and careful management of stakeholders. We can look to the Six Sigma community for better practices in each of these areas. I hope you enjoy reading this book as much as I enjoyed writing it.

MICHAEL: I spent the summer of 1995 working as a bus-driving tour guide in Glacier National Park, using my off-hours for camping and hiking in what is one of the most ecologically interesting and sensitive areas of the United States. When I meet people in the environmental field, they often talk about the glaciers and how they hope to visit the park before the glaciers are gone. The park will always be beautiful, but it's hard to ignore the symbolism of the shrinking glaciers. My time in Glacier, and now enjoying other outdoor activities, is a key source of my motivation to use my work time for good. I have no doubt that as my young sons grow, they will become increasingly demanding of generations before them to demonstrate good stewardship of their inheritance.

My job at Jones Lang LaSalle is to help our clients, usually very large companies, to make their real estate portfolios sustainable. Many people in our industry regard sustainability as energy efficiency. That's wrong. A corporate real estate portfolio is a complicated system of buildings, employee services, local and global suppliers, and materials. We need to consider the economic, social, and environmental impact of the entire system if we truly desire to achieve a sustainable balance, that is, a state where we're not sacrificing the resources of tomorrow in favor of the operational needs of today.

This particular journey began relatively recently for me, in 2007. For some reason, companies were suddenly asking for help in their journey to become sustainable. And the need was going viral, fed by environmentalists, the media, and investors. A popular Internet search tool shows a significant upward-trending count of searches for the word *sustainability* beginning in 2008, by which time the movie, *An Inconvenient Truth*,[1] and the book, *Green to Gold*,[2] had both been out for two years. By mid-2009, when Walmart announced plans to develop a sustainability index to rate products of all kinds, our company and many others had made stronger formal commitments to improve the environmental impact of our operations. In the facilities industry, energy managers were the new rock stars, almost like Web developers during the dot-com boom. Our energy people were endowed with superpowers to reduce energy use, carbon footprint, and operating expenses, all at one time.

While our energy services organization was busy saving companies millions of dollars and our project-management organization was getting dialed in on the U.S. Green Building Council Leadership in Energy and Environmental Design (LEED), my part of the business largely was positioned to solve for "other." Increasingly, this has come to mean large-scale programs, sometimes with defined problem statements and sometimes not. Since corporate real estate organizations solve for the "low-hanging fruit," that is, issues over which they have direct control, they are moving on to more external issues, such as employee engagement and dialog with landlords about space the company occupies but doesn't own. The project roadmap has been varied.

How should a technology company set its next-generation carbon-reduction goals after having spent years driving energy efficiency in its largest facilities? This company had met and exceeded its prior commitments to the EPA Climate Leaders program. The company wanted to set a new goal with a little more data and rigor so that it would not look like a sandbagger. My team and I interviewed stakeholders looking for treasure, and we did some basic analysis of the primary drivers of the company's energy consumption. The company had a pretty good pool of data. We did the best we could. By the time we addressed this issue a couple of years later, this time for a financial services company, we had better tools and a better process for analyzing not just energy-consumption trends but also the impact of future business scenarios that included moving energy-intense data centers, possible acquisitions of other companies, and a better differentiated range of initiatives.

One of our next client opportunities was an energy company on a growth curve in a dozen markets around the world. Facing pressure to build green, how green should the company build? Did LEED make sense as a global construction standard? If so, would building to a Silver certification level be worth the investment premium? And where in these markets might the company find the implementation of new regulatory standards for green building and high penalties for noncompliance? At the time, green building was a mature market priority in Australian markets, thanks to a culture of conservation owing to a lack of abundant natural resources and to efforts tied back to the 2000 Sydney Olympics that emphasized sustainable development. Continental Europe was moderate to strong on green building standards compared with Australia and the United Kingdom. Africa was particularly weak. North America and parts of Asia were somewhere in the middle, according to our analysis of public opinion, adoption of standards, regulatory influence, and cost of green building materials. The analysis was done to support the development of a global green building policy for the company, where the company could adjust targets based on market conditions.

Other projects followed, all interesting in their own ways. Companies wanted to start sustainability programs, scale existing programs, or just find ways to collect good data or train their people.

As a Six Sigma practitioner for years, I was programmed to think about customer-driven improvements (the customers of my clients, in this case). Black belts know that a certain sequence of problem-solving steps—translating customer expectations to metrics to design, assessing for risk, and looking for ways to institutionalize improvements—was the best approach for most business problems. We know the difference between a process-correction need and a change-management need.

It's impossible to be in the sustainability field very long without achieving an authentic motivation for work. In business school, I didn't see a connection between business and environmental issues. Even though I did a lot of camping as a boy and guided tours for a summer at Glacier National Park as an adult, I never saw a connection between the beauty and richness of natural resources and the intellectual and financial pursuits of the business world. But this was only out of ignorance. I've spent countless hours in the past few years opening my mind, exploring the connections, and learning to devote my work life to caring about the environmental, social, and economic issues intertwined in our world. The fact that I'm lucky enough to have this alignment between work and life is a source of strength in both. And I truly believe that the discipline of a Six Sigma approach to business is just the right formula for achieving breakthrough business results and progressing to a sustainable future.

DAN: As an avid outdoorsman who grew up camping, canoeing, and backpacking, I have always had this nagging feeling that my chosen profession in the real estate industry was in conflict with my love of nature. After all, aren't sprawling suburban developments not only eating away at our open lands but also consuming valuable natural resources and contributing to air and water pollution? In my over 30 years in the industry, I have had various opportunities to use my engineering background to focus on improving the energy efficiency of buildings. I was involved in the development of our energy services business at Jones Lang LaSalle, but I always knew that there was a way to have a much bigger impact.

When the CEO approached me about expanding the scope of our energy services practice to consider the broader topic of environmental sustainability, I knew immediately that this was a great opportunity to resolve my inner dissonance and reduce the impact that real estate has on the environment. With buildings being up to 40 percent of the problem, and Jones Lang LaSalle's clients worldwide having large portfolios of properties and ambitious goals to reduce their environmental footprint, it was clear that we could play a significant role. We could make a difference for our clients, the industry, and the environment by offering a broad range of services to help real estate owners and occupiers reduce their environmental impact.

So, with a broader, global perspective in mind, my team and I developed a set of services to address all the real estate decisions that might have an impact on the environment, beginning with the first and fundamental decision regarding the need for more office, retail, or industrial space. We knew that through revised office and occupancy standards, more efficient workstations, or worker mobility programs, many of our clients often could find ways to handle growth within their existing footprint. Or, at the very least, they could minimize the need for new space. Next are decisions about where to locate: close to public transportation or close to where employees live? Then there come decisions about the actual design of the space or building, which can have a dramatic impact on material requirements, waste, and ongoing energy consumption. And finally, building operating practices need to be implemented and maintained that minimize ongoing energy consumption, water consumption, and waste.

We often find that our clients have bits and pieces of these programs built into their real estate processes. In many cases, however, they have what I call good intentions and a collection of good ideas, yet they lack the overall program to drive ongoing success. This book provides the framework for a comprehensive, integrated program that will drive toward the achievement of measurable goals to minimize environmental impact. And I can't think of a better goal.

Thomas McCarty
Michael Jordan
Daniel Probst

Notes

1. *An Inconvenient Truth*, directed by Davis Guggenheim, 2006. The movie is about former U.S. Vice President Al Gore's campaign to educate citizens about global warming.
2. Daniel Esty and Andrew Winston, *Green to Gold: How Smart Companies Use Environmental Strategy to Innovate, Create Value, and Build Competitive Advantage* (New Haven, CT: Yale University Press, 2006).

INTRODUCTION

As Larry Bossidy and Ram Charan wrote in their book about getting things done, "Execution is not only the biggest issue facing business today; it is something nobody has explained satisfactorily. . . . Execution is a specific set of behaviors and techniques that companies need to master in order to have competitive advantage."[1] We feel that the concepts in this book provide content to the issue of developing and executing corporate sustainability programs.

The theme of this book is using the power of Six Sigma to solve the current global challenge of environmental sustainability. One of the most complex problems that organizations face today is achieving success through strategies that are compatible with and supportive of environmental sustainability. At the heart of this problem are the strategies that organizations pursue to manage the people who do the work and the real estate that supports those people. Key to building successful environmental sustainability initiatives is minimizing the negative impacts and improving the positive impacts that workers and buildings have on the environment.

Our intention is to show how typical Six Sigma define, measure, analyze, improve, and control (DMAIC) structures such as program governance, project charters, transfer functions, measurement systems, risk assessment, and process design lend themselves to environmental sustainability. We will use examples of how specific sustainability problems in areas such as carbon emissions, energy conservation, materials recycling, water use, and finance can be solved using Six Sigma tools. The goal of this book is to address key concepts that apply to all businesses. This is a business book, not a technical one. As such, we focus on examples more common to services companies and office buildings than to direct-emitter companies and manufacturing plants. In keeping with our Six Sigma framework, all good initiatives start with a valid business case.

The Six Sigma methodology, as it has evolved over the past 20 years, provides a proven framework for problem solving and organizational leadership and enables leaders and practitioners to employ new ways of understanding and solving their sustainability problems. While business leaders now understand the importance of environmental sustainability to both profitability and customer satisfaction, few are able to translate good intentions into concrete, measurable improvement programs. Increasingly, these leaders are looking to their corps of Six Sigma black belts and green belts to deliver innovative programs that simultaneously reduce carbon emissions and provide large cost savings.

We feel that Six Sigma is a powerful execution engine and that sustainability programs are in need of this operational approach and discipline. Sustainability can't just be taken on because it's got a positive psychic value to influential executives or to employee groups. Six Sigma rigors will help you, as a business leader, to design your sustainability program for both short- and long-term value creation. This book is intended to help you to create value across a multitude of dimensions. Ultimately, it is up to you to decide where you want to lead and innovate and whether sustainability is one of those areas for your company. If it is, we think that this book will help you.

In Chapter 1 we explain concepts for developing the business case for sustainability at the organization level as the first challenge to the corporate executive interested in deploying an environmental sustainability program. The process starts when a senior manager recognizes new influential forces (often external from shareholder or nongovernmental organizations). With a toehold in cost reduction, the company explores other value drivers as well. How does stakeholder management work? What data are needed to get started? We'll establish the connection between corporate demand for natural resources and the effects of this demand on the environment and how to fuse these environmental challenges with strategies to create business value.

In Chapter 2 we share lessons learned from the Six Sigma community about the impact that leadership and management can have on a sustainability effort. It is our experience that leaders find themselves in a paradox of knowing that they *need* to implement a sustainability strategy but find their organization unable to execute because it fails to understand or appreciate the complexity of the organization and therefore the need for a more robust leadership and governance framework that will drive a sustainability plan through to full execution.

The Six Sigma practitioner community has encountered this leadership dilemma since the days when each practitioner completed his or her first analysis and proposed his or her first set of improvements to a process or an organization. Every process change has some level of impact on individuals and organizations. For this reason, every improvement project requires careful attention to the leadership and change-management implications associated with the proposed improvement. Consequently, methodology has developed within the overall Six Sigma framework that enables leaders to determine their improvement targets more strategically and to deploy their improvement projects more effectively. We believe that this Six Sigma management framework can be deployed to enable sustainability initiatives to be more successful. Our premise in Chapter 2 is that leaders can improve their ability to execute on sustainability initiatives if they develop a deeper understanding of the new complexity of the organization that they lead and then tap into a Six Sigma leadership framework to help drive execution through that complex organization.

Chapter 3 deconstructs the major components of a broad sustainability transfer function. We look at the primary drivers of carbon emissions and the use of energy, water, and materials. Six Sigma process improvement relies on establishing the transfer function for desired business results. The examples in this chapter should help you to get started in developing the transfer function for sustainability in your business, identifying areas that combine business importance and underperformance as ripe for chartering improvement projects.

Chapter 4 focuses on describing a variety of international environmental sustainability measurement standards and reporting protocols that have emerged along with related regulatory and voluntary reporting programs. These reporting and measurement standards, both voluntary and regulated, will need to be incorporated into the "big *Y*'s" of any corporate environmental sustainability program.

In Chapter 5 we explore two important factors in driving sustainability initiatives: the power of teams and the importance of effective change management. As we review sustainability initiatives that have failed to achieve their desired results, two distinct but closely aligned factors stand out. They are the failure to plan or execute an effective change-management strategy appropriately and the failure to leverage the use of teams as a part of that overall change-management strategy. We find this fact to be particularly puzzling because so much has been written about the importance of change-management strategy and the overall impact that teams can have in accelerating change. Much can be drawn from our overall Six Sigma body of knowledge to understand how to design and execute a change-management strategy and how to leverage teams in the acceleration of change.

Also in this chapter we review our findings of why so many green project teams have failed to make an impact. We then compare those failure modes with the structures and processes that have enabled success in many Six Sigma program implementations. We explain how this Six Sigma high-performance-teams model fits into an overall change-acceleration strategy to support sustainability initiatives. Finally, we apply these principles to driving positive change across an organization and across all the stakeholders in a sustainability initiative.

Chapter 6 examines the connection between managing buildings and sustainability. Because buildings use so much electricity and water, for example, the ability to manage facilities systems is an important tool in reducing the impact a company has on the environment. We review specific strategies for green buildings. This operational efficiency, however, is only the first step in making a corporate office portfolio more sustainable. We look at location decisions and at decreasing the need for office space, even in engaging with employees through their interaction with the facilities where they work.

In Chapter 7 we thought it important to take three projects from end to end through DMAIC or DMADV (define, measure, analyze, design, verify) structures. We look at the energy efficiency of facilities, reducing office space footprint, and employing green leases as examples of projects with solid financial business cases as well as environmental benefits. We show the key data points and describe the flow of the projects.

Chapter 8 takes a deeper look at applying Six Sigma design tools to sustainability. We show examples of using the house of quality in a sustainability context by using it to design the corporate sustainability program and as a tool within a DMADV design project. We build a basic house of quality to show design requirements for sustainability as a talent attraction and retention strategy. We discuss scorecards and metrics.

In Chapter 9 we explore the role that stakeholders play in the success of a sustainability initiative. Influencing the opinions and actions of groups that can affect your company's short- and long-term sales and stock price is always tricky. When the topic is your company's performance relative to environmental sustainability and climate change, the number of stakeholders has grown dramatically over the past few years, as have the breadth of issues reviewed and the sophistication of the analysis. Many companies have shifted their goals for stakeholder engagement from risk management and talking only about the good things their company is doing to collaboration and transparency in discussing performance strengths and weaknesses. We draw on lessons learned from the Six Sigma community to share improved methods for analyzing the needs, understanding the requirements, and then managing your various stakeholders effectively and proactively.

Note

1. Larry Bossidy and Ram Charan, *Execution: The Discipline of Getting Things Done* (New York: Crown Business Books, 2002).

ACKNOWLEDGMENTS

We would like to acknowledge the support and patience we got from many people during the course of the writing of this book. In particular, we appreciate the support we got from our wives. Our editor, Laima Szymanski, was a fantastic resource and safety net. And many of our colleagues at Jones Lang LaSalle contributed ideas and inspiration for the concepts we put in this book: Meaghan Farrell, Bob Holtz, Tom Poser, Allison Hoppe, Dana Schneider, John Schinter, Gary Graham, Bruce Sirota, and Doug Gottschalk. We'd also like to thank our clients for pushing their own sustainability programs—and us—to high levels of quality and planetary benefit.

A LETTER FROM THE NEXT GENERATION

Dear Grownups,

Keep working on the Earth so when I grow up it will be in good shape.

It is important because I don't want the Earth to be all trashy and I don't want to pick up all the trash by myself. And because I like to play with my friends outside. And I don't want my friends to pick up your trash. I also like to go hiking.

When you put trash in a well did you know it can end up in our rivers and pollute our oceans?

Tips for care:

- Do the three R's: reduce, reuse, recycle.
- Be nice to the E (Earth)
- Do not litter because it's putting trash on the Earth's face.
- Have fun.
- I got this from a skateboarding book: Pick up three wherever you go.

I got this list from a song that we have in the car and from my teacher and from books.

My dad has a book about how companies can take care of the Earth. You should read it.

Love,
Cooper, age 5
(as told to Michael Jordan)

CHAPTER 1

Developing the Business Case

The theme of this book is using the power of Six Sigma to solve the current global challenge of environmental sustainability. One of the most complex problems that organizations face today is achieving success through strategies that are compatible with and supportive of environmental sustainability. At the heart of this problem are the strategies that organizations pursue to manage the people who do the work and the real estate that supports those people. Key to building successful environmental sustainability initiatives is minimizing the negative impacts and improving the positive impacts that workers and buildings have on the environment.

Our intention is to show how typical Six Sigma define, measure, analyze, improve, and control (DMAIC) structures, such as program governance, project charters, transfer functions, measurement systems, risk assessment, and process design, lend themselves to environmental sustainability. We will use examples of how specific sustainability problems in areas such as carbon emissions, energy conservation, materials recycling, water use, and finance can be solved using Six Sigma tools. The goal of this book is to address key concepts that apply to all businesses. This is a business book, not a technical one. As such, we focus on examples more common to services companies and office buildings than to direct-emitter companies and manufacturing plants. In keeping with our Six Sigma framework, all good initiatives start with a valid business case.

Before we address developing the business case for corporate environmental sustainability programs, we will briefly scan the market context within which the investment decision is made.

At the macroeconomic level, policymakers are concerned with issues such as national security, job creation, and consumer price pressure. In this context, social-political efforts to implement sustainable environmental policies gain traction by tying U.S. reliance on foreign oil to threats to national security but lose traction advocating for carbon pricing that could make households pay more for energy. To advance national climate-change legislation, all these issues need to be addressed. The interdependencies of these moving parts make this a difficult challenge. Organizations such as Ceres and Business for Innovative Climate and Energy Policy (Bicep) feel that putting a price on carbon is a change that would greatly increase investment and innovation in "decarbonizing" our economy. Carbon emissions are seen as an economic negative externality that gives emitters a free ride for their impact on climate change. Only by internalizing these costs, some say, will companies be incented to improve. If emissions stay the same and a price is put on those emissions, companies and their customers will have to pay more for the same levels of energy-intensive economic activity. If national economic growth in places such as China required cheap but dirty energy from coal and a price is put on emissions, China would feel that its growth could come only at a price higher than the developed world already enjoyed. The interlocking issues within economies, between nations, and across ecosystems can make one feel that unless all players agree to do something at once, the global community is at a stalemate. How to solve this prisoners' dilemma?

The fact remains that as politically and economically complex as these tradeoffs are, natural resources such as the planet's supply of water are finite, and the Earth's biosphere is a closed system.

Who Should Address This Challenge?

Responsibility for solving our biggest challenges lies with our biggest institutions: government, media, business, and academia, for example. Before the organizations that comprise these institutions will act, they need a compelling reason. The reason might start with a person. In environmental sustainability, as in other fundamental challenges, such as wellness, action can be catalyzed by the personal interest of influential people. (Organizations are, after all, composed of individuals.) People with positional power—a corporate director, a CEO, a senior manager, a manager of an investment fund—might have a personal passion for the environment, and the next thing you know, the company is moving. People with less positional power but who wield influence in the company—a guru engineer, a scientist, a successful brand manager—also might possess a passion to push for change.

Passion is enough to get things moving inside a company. But be careful. As Lauralee Martin, chief financial officer (CFO) of global financial services at real estate company Jones Lang LaSalle, says, "Passion—strong, barely controllable emotions—can easily distract impactful change; but, if focused, it can power success. The Passionate will grab at symbolic solutions, even if ineffective." In other words, don't use passion as the fuel for your corporate sustainability program; use passion to fuel the efforts to develop the business case that generates shareholder value.

Passion can be an important ingredient in getting started because it creates a commitment to innovation and a desire to see new ideas become adopted successfully. But not all business leaders are looking at sustainability because they are passionate about the environment. Some businesses, particularly those in industries with high levels of direct environmental impact, such as mining, oil and gas, chemicals, forestry, and waste, have had environmental programs for decades. At the very least, these programs have developed as a compliance response to regulation. Leading companies have gone beyond compliance to develop environment and safety programs that have become part of the company culture. Recently, businesses also have been prompted by their customers to adopt stronger environmental practices. For example, large companies such as Walmart,[1] Procter & Gamble, and Kaiser Permanente have adopted supplier guidelines for sustainable practices and score their suppliers for progress made on efforts to reduce carbon emissions. In 2000, the Carbon Disclosure Project[2] was launched as a centrally organized effort to get companies to be transparent about carbon emissions, and by the end of 2009, almost 2,500 companies were participating. In 2010, the U.S. Securities and Exchange Commission issued guidance[3] to public companies saying that they should explain the impacts of climate change and climate regulation on their financial disclosure forms. Whether the initial triggers are intrinsic or extrinsic, there are a multitude of triggers that compel a company dialog to consider launching a formal environmental sustainability program.

Once a company commits to a discovery process for a sustainability program, one or more leaders will steward the sustainability agenda across the company. The best way to channel the passion of these leaders is into the development of a business case that suits that company's industry and value chain. In the following two sections we will address development of the business case for sustainability at the company level as well as the business case for individual sustainability projects.

Developing the Sustainability Business Case at the Company Level

When corporate executives feel that sustainability is closer to philanthropy than to core business issues—and many do—they take very little action in this area until an external force triggers the need. In one high-tech company's example, this external trigger arose when a large Europe-based customer demanded that the company put in place a program to measure, disclose, and reduce its carbon footprint. The high-tech company was in no position to refuse, and besides, the cost to measure the company's carbon footprint wasn't very much compared with the revenue from this customer. If your customer hands you lemons, make lemonade! This high-tech company went on to make its products more energy efficient, set and meet multiyear carbon-footprint-reduction goals in a very public way, and become a voice in the global discussion about climate change. Stakeholders and shareholders like it when a company can exhibit the discipline to say what it's going to do and then do it.

Another company we know, a regional bank, was on the receiving end of a shareholder resolution for disclosing its carbon footprint. (Shareholder resolutions on this topic have become more and more common over the past several years, as shown by Figure 1-1.)

In this particular bank, the resolution was a bit of a wakeup call but didn't resonate strongly with senior managers (at least not far beyond the CEO, some of his direct staff, and the Investor Relations Department). Management wasn't hearing from customers that this was very important to them. There have been some stutter-start attempts to explore the effort involved in a carbon-footprint baseline, but the initiative is languishing because the company leaders are not interested enough to figure out the business case. The bank is making good progress on energy-efficiency efforts because of the direct cost savings, but taking the incremental next step in converting kilowatts to carbon, the sustainability equivalent of a "while you're in there" project, has not been deemed worth the cost.

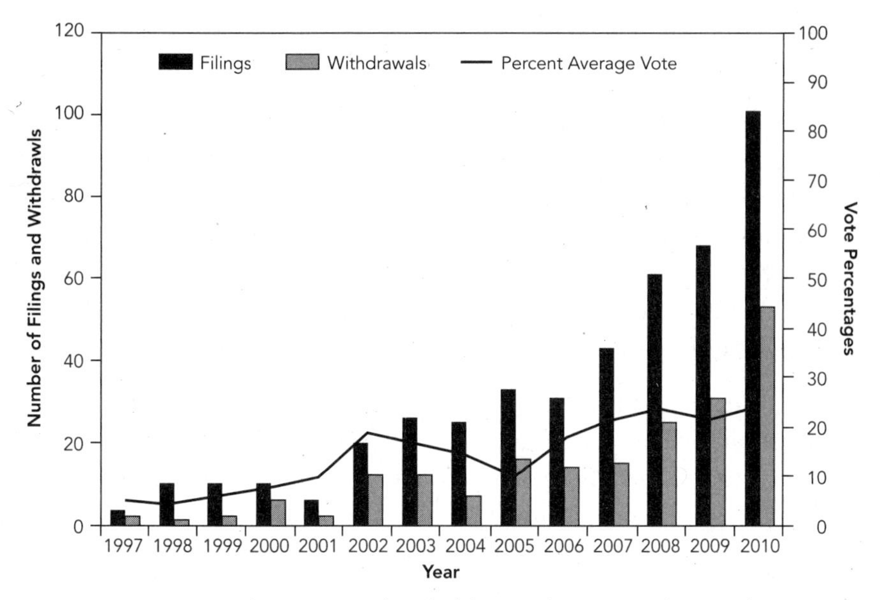

Figure 1-1 North American shareholder resolutions on climate change. (Ceres/Bicep)

For the purpose of illustrating certain examples, we'll be referring to a composite case study from a company we'll call Apex. As an insurance company, Apex is not a heavy emitter of carbon, does not use large quantities of natural resources, nor put an extraordinary amount of hazardous waste into the physical environment. Because of the personal interest—passion, even—of a senior leader, though, Apex initiated a journey into sustainability. And because of the way the company went about it, we suggest that it is an excellent example of building a business case for a sustainability program. Our Apex vice president had plenty of access to the most senior leaders of this company. And he'd been at the company a long time, enjoying a stellar reputation. This vice president could have used his access and relationships to sway other leaders toward a greener future and probably would have seen some success in doing so. But that's not what he did.

First, this Apex vice president started with the basics. He formed a hypothesis that improving the environmental performance of the company was important for Apex to compete as a business. The hypothesis needed to be tested. There were three influential factors that tugged at Apex to go in a greener direction, but no one knew if the factors were strong enough to support a real business case. The factors were

- *Workforce—the interest of current and potential employees.* How strongly did they feel about working for a company that is friendly to the environment?
- *Customers—the interest of clients and potential clients.* To what degree were they incorporating green into purchase decisions?
- *Regulations—the constraints and opportunities afforded by government regulation.* How might Apex be affected by emerging regulations?

Of course, the Apex vice president did not expect to uncover perfect information in all these areas. He wasn't interested in analyzing the question into perpetuity either. But he knew that without the support of a business case, the Apex sustainability program, at best, would blindly follow whatever idea looked popular in the market and, at worst, become a short-lived fad that would sap the company's credibility. It was critical to define the *why*. It took two years of fact collection and microtests to prove the hypothesis and define the Apex sustainability program roadmap in a way that other company executives could understand and support.

Workforce

The Apex vice president, with help from a staff member or two, reviewed public data about the preferences of workers in America. They found research that showed that from 1995 to 2008, the portion of American workers who were either interested in or committed to working for a green company had grown to a majority. Some employees, especially workers under age 35, considered the ecofriendliness of prospective employers in evaluating job offers. Various surveys from employment Web sites and from the Society of Human Resource Managers showed healthy percentages of green preference in new job selections and in staying at a company. By 2009, 600 schools had signed on to the American College and University President's Climate Change Commitment, which covers not only reducing the direct environmental impact of campus operations but also requires integrating environmental sustainability into the curriculum. Current and future workers were being educated on sustainability and finding it important.

Customers

It wasn't very hard for the Apex vice president to find evidence that customers expected Apex to show evidence of good environmental practices. By reviewing requests for proposals (RFPs) from potential customers, the vice president was able to establish that some (but not all) customers were asking for information about Apex environmental practices. Information requests ranged from questions about Apex's initiatives, policies, and participation in third-party organizations to questions about how Apex could improve the customer's environmental impact.

The following list represents some of the questions our clients and we see in RFPs:

- Please detail your corporate environmental goals and policy.
- What awards and recognition has your company received for environmental performance? What certifications has your company attained?
- Describe the policies and procedures that your company uses to ensure minimum impact on the environment.
- Describe your company's commitment to international sustainability frameworks such as the Hanover Principles.
- Is your company participating in the Carbon Disclosure Project, Global Reporting Initiative, or similar efforts?
- How is environmental performance incented and governed at the board of directors and/or at the top executive level in your company?
- How does your company account for greenhouse gas emissions?
- How does your company develop (and maintain) energy policies, incentives, and awareness programs to encourage energy conservation?
- How can your company help our company identify opportunities for tax incentives or utility company rebates?
- What processes has your company certified using ISO 14000?

Apex knew that to seek buy-in from executives in the company, a laundry list of RFP questions would not be very effective. Instead, the Apex vice president summarized his findings in a simple profile of "green customer expectations" so that these features would be the foil for future discussions. The green customer expects Apex to:

- Demonstrate executive commitment to sustainability, backed up by policy

- Report on environmental goals and progress (most likely in areas such as carbon footprint)
- Manage sustainability up and down the supply chain
- Have a climate-change mitigation strategy
- Participate in external organizations committed to mitigating climate change or to helping the environment

In presenting this profile, the Apex vice president was transparent that this represented a small proportion of Apex customers. Backed by research, however, he now knew what to look for to determine if this was a growing trend, not to mention that the list of customer expectations helped to form design criteria for a sustainability program.

Regulations

In addition to looking at trends in local, state, and federal regulations, it's a good idea to identify the regulatory bodies that affect your company and industry. Since Apex is an insurance company, one important such body is the National Association of Insurance Commissioners (NAIC). The NAIC recently declared its intent to require insurers to submit an annual climate-risk disclosure survey starting in 2010. This issue of disclosing climate risk was getting traction in the market, not just in insurance. As a directive, disclosing risks seems straightforward. In order to do this, though, companies must identify scenarios where they could be exposed to energy supply fluctuations, flooding, changes in weather patterns, and the like. Some of these factors could pose risks upstream in the company's supply chain (think of the potential impact to Starbucks or to Levi Strauss of drought in countries that produce coffee or cotton). Because these issues can be very localized, a company doing business in markets across the United States or around the world has its work cut out for it.

By reviewing information from sources such as the U.S. Green Building Council, the Pew Center on Global Climate Change, and the Council of State Governments, Apex discovered that 42 states had passed new legislation or had pending legislation regarding green building codes, greenhouse gas, and landfill restrictions. Although some municipal codes offer opportunities such as expedited building permits for green building commitments, most new regulations introduced additional constraints on company operations or additional costs for monitoring and reporting. As a services business, Apex wasn't accustomed to environment-related regulations, as would be a manufacturing or chemicals or mining company. (In these latter companies, leadership on sustainability programs often emerges from the environmental health and safety (EH&S) organization. However, EH&S groups also tend to possess a strong compliance mentality, which can make it difficult for companies to see sustainability as an opportunity for innovation and growth.)

Developing a focused business case—from the outside in—becomes part of an iterative process. Because there are so many stakeholders who care about the company's environmental and social performance, formal processes to engage stakeholders have become a de facto part of sustainability programs. At the early stages of setting up the business case, we recommend not overengaging with stakeholders. Too much input can become noise, and too many requests can become difficult to prioritize. However, it is important to engage company decision makers to get buy-in for launching the program. Make the case succinctly, and propose some simple initiatives that have a clear return on investment (ROI). Taking action on the first initiatives will build further support for the program and earn the right to expand scope and scale—and the funds generated from early success can be used to launch next-generation initiatives. We work with one energy company that launched its internal energy-efficiency program with a pool of capital dollars for funding energy conservation projects. The pool is used as an internal investment program with its own payback criteria. Projects that are funded from the pool reimburse the fund, which, as it grows larger, can fund more projects.

Matching centralized funding with centralized expertise and project-approval oversight can be a powerful tool for ensuring that the right projects get done and for aggregating the results. When the company is very large and/or has a distributed supply chain including many partners, the central pool of funds and expertise may be the only way to give proper attention to complex projects such as renewable power generation. But centralized funding is not just for hard-dollar-return projects. We also have seen companies that use a central pool of funds for employee volunteer projects. At Intel Corporation, for example, through its Sustainability in Action program, the company invites employees to apply for project funding of "passion projects" that make a difference in the community. In 2009, Intel funded nine employee volunteer projects to encourage innovation in environmental community service, including a community recycling project in Russia, environmental education sessions in schools in Ireland, forest cleanup and conservation activities in Oregon, and a city air-quality monitoring project in Arizona.[4]

Over time, the business case that launched your company sustainability program will be augmented by new goals that operationalize the desired benefits. These initiatives will change the company's role in its community and in the market. As new goals are set and the program expands, expect the list of spectators to grow as well. (In fact, some companies avoid building a robust sustainability program because they do not want to attract attention from stakeholders that may use negative publicity in support of a cause, but in a way that harms the company. Fortunately, many nongovernmental organizations have realized that it is in their best interests to be engaged and constructive.) Engaging stakeholders of different types can generate ideas and future initiatives. And even if your company chooses not to invest in the programs desired by stakeholders, simple engagement can have benefits in and of itself. We deal with the issue of stakeholder management in Chapter 9.

In their book, *The Balanced Scorecard: Translating Strategy Into Action*,[5] Robert Kaplan and David Norton suggest that effective corporate strategies are built from four perspectives:

1. *The financial perspective.* To succeed financially, how should we appear to our shareholders?
2. *The customer perspective.* To achieve our vision, how should we appear to our customers?
3. *The internal business process perspective.* To satisfy our customers, at what business processes must we excel?
4. *The learning and growth perspective.* How do we build capability to accomplish this plan?

Categorizing the research collected during the initial discovery phase into these areas generates important discussion among the participants in the sustainability effort. By setting quantifiable, overarching goals in each area, the team can show how the proposed sustainability program would add value to the company's performance. Future projects then can be focused on solving for the quantified goals.

In 2007, Procter & Gamble announced a series of sustainability goals, including one to "develop and market at least $50 billion in cumulative sales of sustainable innovation products" by 2012.[6] This is a strong statement to shareholders about how sustainability will be part of the company's financial success.

Method, a San Francisco–based consumer products company, was started with a clear mission in the customer perspective category: "To make products that work, for you and for the planet, ones that are as easy on the eyes as they are on the nose."[7] The theory is that customers who value these principles will gravitate toward Method products (and that talented people who believe in these values will want to work at Method).

Once a company decides to adopt a sustainability program, many start with internal business processes because these processes are straightforward to control. In many cases, companies that want to sell green products and services also feel that it is important to first get their internal house in order. It is common to see first-generation initiatives to reduce energy and water consumption as well as waste generation. Two of Walmart's environmental goals are in this area: To be supplied 100 percent by renewable energy and to create zero waste.[8] By looking at processes such as production, transportation, and facilities maintenance, companies can make great progress in reducing the environmental impact of internal operations.

Developing corporate strategy with learning and growth in mind is not a common explicit goal. In some ways, though, this can be the most important investment a company makes because without it, the company likely will find itself moving resources around, reacting to trends, and underfunding the business volume needed to create revenue. A company can't expect to build new sustainable products if it doesn't create new skills and knowledge among its product-development organization. A company can't expect to reduce energy use if it doesn't help its employees to know what to look for. New technologies that help companies track and report carbon emissions are important capabilities builders. A recent study[9] by GreenBiz.com and Groom Energy estimates that worldwide unit sales of enterprise carbon accounting software are rising from 50 in 2009 to a projected 1,500 in 2011. Companies are building capabilities at an accelerated pace.

As company objectives are set in these strategic areas, Six Sigma teams can be mobilized to make progress across the company. Some projects just should be done—not every idea needs to be taken through the DMAIC process. Where there is an element of complexity, where discovery is needed to uncover critical success factors, and where there is potential to affect the customer experience, though, black belts should seek to develop business cases for sustainability projects.

Developing the Sustainability Business Case at the Project Level

Consistent with Six Sigma improvement projects in other functional areas, the best sustainability projects will be in alignment with company priorities and will be driven by data. In the project charter, the business case is a summary of the project rationale and a description of the clear line of sight to top-level drivers of business value. Using performance data to select projects likely will be a challenge if the company sustainability program is new (see Chapter 4). However, collecting a reasonable sample of data and facts is the best way to support the project proposal and create a basis for discussing the merits of the project over other potential investments when presented for approval. Keep in mind that projects to improve the company's sustainability performance will compete for resources with projects for other purposes.

It can be particularly difficult to get approval for operational-efficiency projects—even projects with good payback periods—when the same resources could be allocated to projects that expand production capacity. On the other hand, if your company is a consumer-products business with a strong emphasis on brand, projects with weak hard-dollar-payback periods might be strengthened if the anticipated outcomes have good public relations value. We have seen this go both ways—at one company, investments in a new solar array were approved in part because of the reputational value of such a high-profile project. At another company, the business case for proposed capital improvements to improve energy efficiency to attention-getting levels could not include public relations value because the company would not publicize results not tied to achievements in *product* performance.

Leading marketing companies have divergent opinions about the degree to which consumers in the United States care about how green a product is.[10] Whereas Frito Lay promotes the fact that some Sun Chips are made in a plant[11] that is powered by solar energy that's better for the planet, Proctor & Gamble promotes Tide Coldwater[12] as a product that saves consumers money by reducing the energy required to wash clothes without being explicit about benefits to the planet.

It can be difficult to build the business case for a sustainability project solely on the merits of an improvement in company environmental impact (unless it is to comply with government regulations). For example, we worked with senior managers at one company who were interested in projects that would be inspirational to employees. One of the options explored was to reduce food waste from company cafeterias. In the United States, a large portion of food is discarded.[13] When organic material ends up in a landfill, it decomposes and gives off methane,[14] a greenhouse gas, 21 times more insulating in the atmosphere than carbon dioxide. From an environmental standpoint, reductions in this area would be important. From a business-value standpoint, though, the case was difficult to make. Company managers had the data to show that food waste was an important issue, and they even knew other companies that had addressed it successfully, but the business case based on reduced waste hauling costs could not win approval without the extrapolated financial value of improved landfill diversion rates and improved employee relations. These latter values were too abstract compared with the well-defined costs of food dehydrator equipment and other resources. We suggest that there is a simple Venn diagram at play here (Figure 1-2).

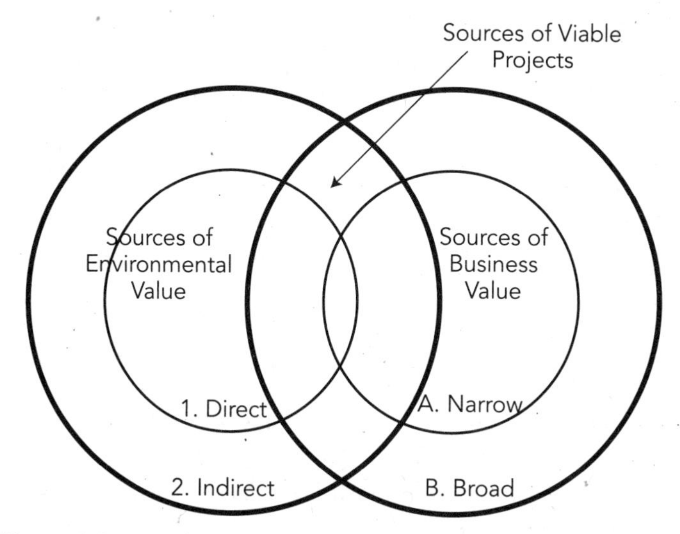

Figure 1-2 Overlapping environmental value and business value.

In this Venn diagram, one circle represents the set of drivers of environmental value; the other represents the set of drivers of business value. Of all the possible sources of environmental value (e.g., carbon footprint, pollution, extraction, contamination, deforestation, and the like), the areas where the company makes a direct impact are part of the inner circle (1-Direct), whereas others are part of the larger circle (2-Indirect). As we discuss in Chapter 3 and in other parts of this book, for most companies, electricity use by itself results in an indirect environmental impact but receives a lot of attention because energy conservation is a direct source of business value (cost savings).

The direct environmental impact of electricity use happens at the source of power generation, and if fossil fuels are burned, one category of impact is carbon emissions. In our Venn diagram, a project that reduces electricity use has an indirect environmental benefit (reduced carbon emissions) and a direct business value (operating expense reduction from using less electricity). Reduction of paper use could be a source of direct environmental value (reducing waste to landfill) as well as a source of indirect environmental value (reduced energy consumption and reduced harvesting of trees) and a source of business value by reducing money spent purchasing and disposing of paper.

If your company is interested in sustainability only as it pertains to cost savings, the set of business-value sources is relatively small (A-Narrow). If the company uses sustainability as a driver of product sales or brand value, the circle is larger (B-Broad). In any case, sustainability projects will not likely be approved if they are not conceived from the overlap between the two circles. The company sustainability lead might consider it his or her job to create as many connections as possible between environmental value and business value, making the circles as large as possible for their company. Sources of business value at your company might include

- Revenue growth
- Expense reduction
- Cost of capital
- Brand-recognition improvement among consumers
- Improved reputation among stakeholder groups
- Development of a leadership position (e.g., setting direction or pace in the company's industry)

- Market expansion abilities (by geography and/or by industry type)
- Improved speed to market
- Product quality improvement (e.g., reduction in environment-related defects, realization of environmental opportunities)
- Risk management
- Company culture and/or improving employee retention
- Ability to attract new talent

A special area of consideration is reserved for valuing negative externalities. In this area, the business case is made by putting a financial value to impacts that otherwise sit in the Environmental Sources of Value circle. The most common way this happens is by purchasing carbon credits. Companies that have committed to becoming carbon neutral will work to become as energy efficient as possible but then must either buy or build a supply of renewable carbon-free power or purchase carbon credits to offset the carbon emissions from energy used by the company. The cost of the credits amounts to a sort of self-imposed tax on the organization. Therefore, any improvement projects that reduce energy consumption from fossil fuels also will reduce the "tax." For companies that purchase carbon credits, reduction in these costs should be included in the business case for energy reduction–related projects.

The leaders of the sustainability program define these boundaries, get executive buy-in for the boundaries, and activate Six Sigma teams against the right set of opportunities. When drafting the charter for an improvement project, the business-case section should include the direct/indirect and narrow/broad value drivers that are relevant to the company's sustainability goals.

The business case for an individual Six Sigma project should tie back to the business case of the enterprise-wide sustainability program and its reflection of the company's values and financial statements. Don't ignore competitive issues. If your company is looking for a leadership position in sustainability, you might consider your relative placement on rating systems such as the Carbon Disclosure Project, the Newsweek Green Rankings list, or the JustMeans Global 1000. If your corporate sustainability program uses these systems as a scorecard or uses membership in the Dow Jones Sustainability Index as a goal, then projects that improve performance against the requirements of these programs should be given fair consideration.

We also recommend researching the accomplishments of other companies to support the opportunity targeted by your project. For example, a large part of the corporate carbon footprint for services companies usually includes corporate travel on airplanes. At one financial services company with which we worked, corporate air travel accounted for about 10 percent of the company's carbon footprint. At another company, Cisco, air travel accounted for just over half the 2006 carbon footprint.[15] Over the coming three years, Cisco would deploy changes to business process, management practices, and culture, along with a portfolio of high-definition video teleconference systems, all aimed at reducing travel by up to 40 percent. The carbon reduction, cost savings, and marketing value for Cisco Telepresence systems added up to millions of dollars. For some companies, the business case for travel reduction might be less desirable or more difficult to realize. But the thought process for multiple connections between environmental and business values applies universally.

Developing the business case for an energy-efficiency project at any company is usually easy because even though the environmental value drivers are indirect, these drivers are part and parcel of the highest-profile issue in sustainability today—climate change through carbon emissions. Energy consumption is also easy for executives and employees at all levels to understand because everyone pays for energy at home. The business-value drivers of energy conservation are usually direct cost savings. And the reduction in carbon footprint has stakeholder reputation value. In order to establish the business case for energy-efficiency projects, the black belt project manager needs to know the current energy consumption volumes and rates (data here are also relatively easy to get) and some idea of the sources of current energy inefficiencies. In office buildings, typical areas for savings include lighting, heating, and cooling systems. Once the black belt confirms the utility rate structure and whether any past improvement projects have been done in these areas, a project business case generally can be made for improving the energy efficiency of a building or group of buildings. Benefits are likely to be cost savings and carbon-footprint reduction.

Another common topic for an improvement project is reducing office waste that goes to landfill. To establish the business case for the project, the black belt looks first to establish that there is a problem worth assigning resources to solve. In the environmental circle, value comes from both impact on land where the waste goes, as well as impact on land that provides the resources to create the material to begin with. There are also indirect environmental impacts because office materials require energy to produce and to transport. Given that trash from the office (e.g., paper, cardboard, toner cartridges) has to go somewhere after it is used, it is easy to see the motivation for preventing the accumulation of materials in landfill. And the more paper that is used in the office, the more natural resources are required to produce that paper. On the business side of the equation, if waste can be reduced, then costs to haul waste away from the office should be reduced, cost to procure paper should be reduced, and employee satisfaction from working for a company that recycles should be improved. Recycling is a top sustainability issue among employees at every company we talk to. There is, however, a lot of variation from company to company in willingness to invest labor and other resources into solving this problem. Once a solution is designed, keep in mind that labor also will be required to bring the solution to scale across a large organization.

Before you limit your thinking to energy-conservation projects or recycling, consider that your colleagues at work are passionate about many more issues than these two. Consider the adage, "It's better to work to live than to live to work." We're seeing more and more singularity between the roles that individuals play as workers, consumers, citizens, and family members. In other words, behind those roles, we are one person, the same person at home and at work. The person who cares about recreation and art and science is the same person who works in the accounting or engineering department. People the world over spend billions of dollars and countless hours engaged in sports and entertainment and going to museums. This suggests that your sustainability program has a chance to make work more active, fun, and creative because that's what people like to do anyway.

Pushing the limits of action to make your company sustainable is an exercise in personal leadership. There are endless models of personal leadership, but consider one good example, *The Leadership Challenge*, by Jim Kouzes and Barry Posner,[16] wherein the authors discuss what they consider the five fundamental practices of leadership:

- Challenge the process.
- Inspire a shared vision.
- Enable others to act.
- Model the way.
- Encourage the heart.

Managing your company's use of natural resources could be approached as an administrative task with many barriers. And changing light bulbs in your office to make it more energy efficient can be personally satisfying. But expanding the circles of value creation for your company, applying a more collaborative approach across functional boundaries, and working up and down the sustainability transfer function to activate other intrinsically motivated individuals is more like a mission and less like a job.

Chapter Summary—Key Points

- Solving for environmental sustainability at the global-systems level requires multiple players to make contemporaneous decisions. Responsibility for solving our biggest challenges lies with our biggest institutions.
- Sustainability strategies should be driven not by passion but from a strong business case—where financial and strategic drivers of shareholder value overlap with drivers of environmental value.
- At the company level, research and develop the business case for initial investments in sustainability by understanding the expectations of your workforce, your customers, and your regulators.
- Every Six Sigma business-improvement project requires a business case. Use performance data that reinforce the shareholder value as well as the environmental value of the project. Collect data and facts in order to shape the business case.
- Keep in mind that even strong sustainability improvement projects will be competing for resources against projects for other purposes, for example, product development.
- Leading companies have divergent views about how strongly consumers can be influenced by environmentally friendly product attributes.
- Assigning a dollar value to metric tons of carbon can assist in evaluating the ROI for sustainability projects.
- Sustainability program leaders should work with other company executives and Six Sigma project champions to establish the rules for how project business cases can be quantified, for example, what is the value of carbon abatement or of improved employee engagement? Or what is the value of exceeding the sustainability performance of your competitors?

- The business case for individual Six Sigma projects should reinforce the business case for the company sustainability program.
- There is a growing trend for singularity in work-life roles: The same people play the role of worker, consumer, citizen, and family member. Your sustainability program has a chance to grow company culture and your ability to attract and retain talent by making work active, fun, and creative.

Notes

1. Walmart President and Chief Executive Mike Duke (July 16, 2009): "Customers want products that are more efficient, that last longer and perform better. And increasingly they want information about the entire life cycle of a product so they can feel good about buying it. They want to know that the materials in the product are safe, that it was made well and that it was produced in a responsible way."
2. Carbon Disclosure Project: www.cdproject.net/enUS/WhatWeDo/Pages/overview.aspx.
3. Securities and Exchange Commission 2010 guidance press release: www.sec.gov/news/press/2010/2010-15.htm; final rule: www.sec.gov/rules/interp/2010/33-9106fr.pdf.
4. Intel Corporation 2009 Corporate Social Responsibility Report, p. 87.
5. Robert S. Kaplan and David P. Norton, *The Balanced Scorecard: Translating Strategy Into Action* (Boston: Harvard Business School Press, 1996).
6. Procter & Gamble announced five strategies related to sustainability: products, operations, social responsibility, employees, and stakeholders. "Sustainable Innovation Products are included if they have launched in market since July 1, 2007, and have a greater than 10 percent reduction in one or more of the following indicators without negatively impacting the overall Sustainability profile of the product: A. Energy, B. Water, C. Transportation, D. Amount of material used in packaging or products, E. Substitution of nonrenewable energy or materials with renewable sources." www.pg.com/en_US/sustainability/strategy_goals_progress.shtml.
7. Method set goals to create home products that are clean, safe, green, well designed, and have appealing fragrance; www.methodhome.com/methodology/our-story/we-are.
8. Walmart's environmental goals are "To be supplied 100 percent by renewable energy; To create zero waste; To sell products that sustain people and the environment"; http://walmartstores.com/sustainability/.
9. Groom Energy and GreenBiz.com, *Enterprise Carbon Accounting: An Analysis of Corporate-Level Greenhouse Gas (GHG) Emission Reporting and a Review of*

Emerging GHG Software Products, January 19, 2010. Groom Energy Solutions and Pure Strategies (Groom Energy: Salem, Massachusetts; Greenbiz.com: Oakland, CA); www.groomenergy.com/files/ECA_Jan11_update_sample_free_excerpt.pdf. GreenBiz.Com:. www.greenbiz.com/product/enterprise-carbon-accounting-jul-2010.

10. Procter & Gamble and Ipsos Public Affairs findings from their 2010 Consumer Conservation Survey: "About 74 percent of consumers say they would switch to another brand if it helped them conserve resources without having to pay more. About 37 percent say the reason they don't lead a more environmentally friendly lifestyle is a lack of enough information about what to do. About 58 percent say they would be at least very likely to change the way they perform daily chores if it helped them reduce waste, save energy and save water at home." www.prnewswire.com/news-releases/pg-launches-initiative-to-make-conservation-of-natural-resources-more-user-friendly-87640202.html.
11. www.sunchips.com/healthier_planet.shtml.
12. www.tide.com/en-US/product/tide-coldwater.jspx.
13. University of Arizona UA News: "According to recent statistics cited by the U.S. Environmental Protection Agency, about 25 percent of food in America is discarded. A 2004 study released by former University of Arizona anthropologist Timothy W. Jones estimated that 40 to 50 percent of all food ready for harvest never gets eaten. Jones' study went on to say that nationwide household food waste alone added up to $43 billion annually. On the environmental side of the equation, Jones estimated that reducing food waste would be beneficial through reduced landfill use, soil depletion, and applications of fertilizers, pesticides, and herbicides." Available at http://uanews.org/node/ 10448.
14. U.S. Environmental Protection Agency: "Methane (CH_4) is a principal component of natural gas. It is also formed and released to the atmosphere by biological processes occurring in anaerobic environments. Once in the atmosphere, methane absorbs terrestrial infrared radiation that would otherwise escape to space. This property can contribute to the warming of the atmosphere, which is why methane is a greenhouse gas. Methane is about 21 times more powerful at warming the atmosphere than carbon dioxide (CO_2) by weight." Available at www.epa.gov/methane/scientific.html.
15. Cisco Announces Carbon Reduction Initiatives, September 21, 2006. Available at http://newsroom.cisco.com/dlls/2006/ts_092106.html.
16. Jim Kouzes and Barry Posner, *The Leadership Challenge*, San Francisco, California; Jossey-Bass Inc. (1995).

CHAPTER 2

Sustainability and the Collaborative Management Model

The Paradox

If leaders honestly buy into the business case for sustainability, then they understand that having a solid commitment to sustainability supported by a robust, operational sustainability plan is a critical factor in the long-term success of their organization. And most leaders surveyed would indeed answer in the affirmative that sustainability is a key factor in the success of their organization. It is paradoxical, then, that so few leaders can show strong evidence that they have put in place a robust, operational sustainability plan. It is our experience that leaders find themselves in this paradox of knowing that they need to implement a sustainability strategy but finding their organization unable to execute because they fail to understand or appreciate the complexity of their organization and therefore the need to have a more robust leadership and governance framework that will drive a sustainability plan through to full execution.

The Six Sigma practitioner community has encountered this leadership dilemma since the times that each practitioner completed his or her first analysis and proposed the first set of improvements to a process or an organization. Every process change has some level of impact on individuals and organizations. For this reason, every improvement project requires careful attention to the leadership and change-management implications associated with the proposed improvement. Consequently, methodology has developed within the overall Six Sigma framework that enables leaders to determine their improvement targets more strategically and deploy their improvement projects more effectively. We believe that this Six Sigma management framework can be deployed to enable sustainability initiatives to be more successful. Our premise in this chapter is that leaders can improve their ability to execute on sustainability initiatives if they develop a deeper understanding of the new complexity of the organization they lead and then tap into a Six Sigma leadership framework to help drive execution through that complex organization.

New Levels of Organizational Complexity

There was a time when most organizations could be viewed as pyramids, with a leader at the top, eager employees at the bottom, and some layers of management and administration in the middle. Organized much like military organizations, these pyramids executed on their mission through a command-and-control leadership model. As we all know, pyramids lost their relevance with the Egyptians some 2000 years ago. They are simply too rigid and nonresponsive to change. Likewise, organizational pyramids are not flexible enough to adapt to the rapid pace of change in the external markets that most organizations face. In addition, the command-and-control pyramid is no longer attractive to workers interested in more flexibility, diversity, and autonomy in their work environment.

In response to the rapidly changing requirements of the external markets, as well as the changing requirements of the workers an organization needs to serve its markets, an organizational structure has emerged that can be viewed as four concentric circles or, more accurately, a four-leaf clover because, just like the plant, the leaves are connected and interdependent on one another to live or thrive (Figure 2-1).

This shamrock design is light-years ahead of the classic organizational structure, the pyramid, for good reason. A pyramid-shaped command-and-control structure doesn't last because things change too fast. The kind of work required changes. The required working relations change. Who the organization needs for suppliers, partners, and strategic sources changes. And leaders want them to change because change is what keeps an organization fluid, responsive, competitive, and alive. The shamrock or concentric-circle design is a response to one simple fact in organizational development—*things change.*

Figure 2-1 The "shamrock" organizational structure.

Today, people work for different reasons and different rewards. People no longer work, you might have noticed, as parts of a machine or as cogs in a wheel. Instead, we increasingly work

- In networks and alliances
- In collaborations
- On projects and initiatives of varying duration
- By choice, according to the circumstances of the person

In response, the shamrock organization is made up of four very different groups of people. They are four distinct groups with different expectations. They are managed differently. They are paid differently. And they are organized differently.

The first leaf is made up of core workers. This is the increasingly small group of people who are essential to the organization. They are essential because they own the organizational knowledge. Hard work and flexibility are demanded of them. These people are highly committed. As a result, they are well paid and incentivized. Senior core people are transformed into partners, rather than employees, with high leverage through performance-based compensation. In order for these people to stay highly committed and strategic, though, they have to be involved in work they love to do. Somebody else must do the work that is not so exciting.

You keep passionate people employed in high-level strategic thinking by having people in the second leaf equally passionate about the tactical, day-to-day work. The second group is made up of strategic supplier alliances. In this new organization, all nonessential work is contracted out—to both individuals and organizations. Because the core is not only smaller, it is also more expensive, and if the work of the second leaf were paid at in-house salaries, the organization would go bankrupt. And more important, you can't ask people in the core to do this work and stay engaged.

Therefore, strategic supplier alliances are the effect of the organization's recognition of what work is crucial to the organization. And while outsourcing to strategic supplier alliances might seem to some to be a relatively recent phenomenon, it has long existed in the form of a contractual fringe. It is a way of exporting uncertainty in inventory and labor. In this group, workers are paid for results, not for time. Workers in the second leaf excel at independently achieving day-to-day results. The role of the core is to specify the results, not to oversee the methods.

The third leaf is a group of subject-matter experts. These subject-matter experts usually are external to the core employees and are focused on highly complex and technical issues that affect the current profitability and future of every firm. They are, generally, external to corporations because they dedicate much of their time to research, writing, and teaching. Typically scientists and research analysts, they are found within the academic community, in industry-supported associations, or within governmental agencies. Regardless of their physical location, firms are highly dependent on these subject-matter experts because they conduct the research and analysis, and they make the discoveries that determine future outcomes. These outcomes are brought on through technological breakthroughs, game-changing innovations, or legislative changes. Forward-thinking firms have found ways to partner with these subject-matter experts in ways that are mutually beneficial.

This leads us to the final element of the shamrock organization. If you get this right, the most significant leaf of your shamrock is the customer. The organization's alignment and integration with the customer are so tight as to be seamless. In this model, there is no longer the sense of "inside" or "outside" the organization.

The shamrock model depicts an organization that is fluid, dynamic, and highly responsive to changing market requirements and the diverse requirements of each individual. In this organization, individuals will move between the groups as their passions, needs, interests, and circumstances evolve..

What we have attempted to describe in the past few pages is the reality and rationale behind the complex organization model that reflects the operating model for most organizations today. The shamrock organization is the current reality for most organizations. This is a fact, whether the leadership recognizes the design or not.

The important implication to implementing sustainability initiatives is that in order to achieve holistic success in sustainability, leaders must inspire and motivate action through four very distinctive organizational constituencies with four very distinctive sets of needs and motivations. Those constituencies are

- Core employees
- Strategic suppliers
- Subject-matter experts
- Customers

The needs and motivations of these constituencies relative to sustainability must be dissected, understood, and then addressed through a leadership and governance model that is likely to be much different from the model in current practice inside most organizations today (Figure 2-2).

In order to address this complex leadership challenge, we have introduced a leadership model that we call the *collaborative management model*, which we will discuss at length later in this chapter. In order to understand the rationale behind the collaborative management model, however, we think that it is useful to review lessons learned from the evolution of Six Sigma as a model for leadership and transformational change.

Constituency	Motivation	Leadership Implications
Core employees	*"I want to work for an organization that is aligned with my values where I can feel engaged and purposeful."*	How to operate in an environmentally friendly manner and engaging employees in the initiative
Supplier alliances	*"I want to maximize my near-term profits while ensuring future growth and profitability."*	How to motivate suppliers to engage in sustainable practices that may reduce their profitability in the short term
Subject-matter experts	*"I place the future of the planet ahead of corporate profits."*	How to align interests with subject-matter experts that have low regard for corporate profitability
Customers	*"I want the highest value at the lowest possible price."*	How to communicate the value of environmentally friendly products while maintaining lowest possible costs

Figure 2-2 Constituency motivations and leadership implications.

Evolution of the Six Sigma Leadership Framework

As we review the history of Six Sigma business-improvement systems—particularly when the subject is breakthrough improvement as measured by financial impact—there is a basic assumption: the notion that a leadership team is leading the way, pointing teams to appropriate targets, and then monitoring improvement efforts to make sure that projects stay on track to achieve the desired financial returns.

In practice, unfortunately, case after case of failed efforts illustrates that despite all the talk about the role of leaders, it is often the lack of cohesive, continuous leadership that is a key driver of failed attempts at business improvement. This is especially true when the reason for failure is the inability to achieve a desired financial impact.

Most leaders understand intuitively that their organization's business-improvement efforts will have a higher likelihood of sustainable success if they remain active in the process. Few have a concrete model for putting this understanding into practice.

A review of successes at organizations such as General Electric, Allied Signal, Honeywell, Caterpillar, Bank of America, and Motorola demonstrates that breakthrough improvements—and the financial gains that are associated with them—occur when senior leadership moves beyond thinking about Six Sigma as a tool set for driving continuous process improvement. Successful champions of the discipline view Six Sigma as an integrated set of leadership and governance practices supported by a specific set of leadership behaviors.

Such leaders share common themes in their practice of a Six Sigma leadership framework:

- A Six Sigma macro model for driving organizational change
- A continuous leadership process for monitoring and driving execution
- An established set of expected leadership behaviors

The Six Sigma leadership and governance framework is a set of activities and tools that enables a leadership team to align on their strategic objectives, establish their critical operational measures, determine their organizational performance drivers, and then use that framework to implement, drive, monitor, and sustain their Six Sigma effort.

Six Sigma as a Leadership Framework

In order to create the context for Six Sigma as a leadership and governance framework, it is useful to understand the evolution of Six Sigma from metric to methodology to management system. Six Sigma has been described as a metric, a methodology, and an overall leadership and governance framework. In essence, it is all three at the same time, and we'll spend a little time describing the evolution of each.

A critical paradigm for leadership teams to adopt is an understanding that Six Sigma as a best practice is more than a set of problem-solving and process-improvement tools. Based on review of successes at the organizations cited earlier, it can be demonstrated that breakthrough improvements occurred when senior leadership of these organizations adopted Six Sigma as a leadership and governance paradigm. At its most conceptual level, Six Sigma could be defined as a business-improvement methodology that focuses tools and management practices on four key areas: understanding and managing customer requirements, aligning key processes to achieve those requirements, using rigorous data analysis to understand and minimize variation in those processes, and finally, driving rapid and sustainable improvement to the business processes. Six Sigma, when practiced at three levels, will lead to increasing levels of impact (Figure 2-3).

At the metric level, Six Sigma practitioners use process metrics such as defects per million opportunities (DPMO) and cycle-time measurements to monitor and manage their processes. For organizations focused on sustainability improvement, carbon dioxide (CO_2) emissions also would be used as a key measurement. At the methodology level, Six Sigma teams use a disciplined methodology often described as the define, measure, analyze, improve, and control (DMAIC) methodology systematically to analyze key processes, determine sources of variation, develop improvement plans, and implement sustainable solutions.

Clearly, the use of a consistent set of metrics can greatly aid an organization in understanding and controlling its key processes. So too the various problem-solving methodologies significantly enhance an organization's ability to drive meaningful improvements and achieve solutions focused on root cause. Unfortunately, experience has demonstrated that good metrics and disciplined methodology are not sufficient for organizations that desire breakthrough improvements and results that are sustainable over time.

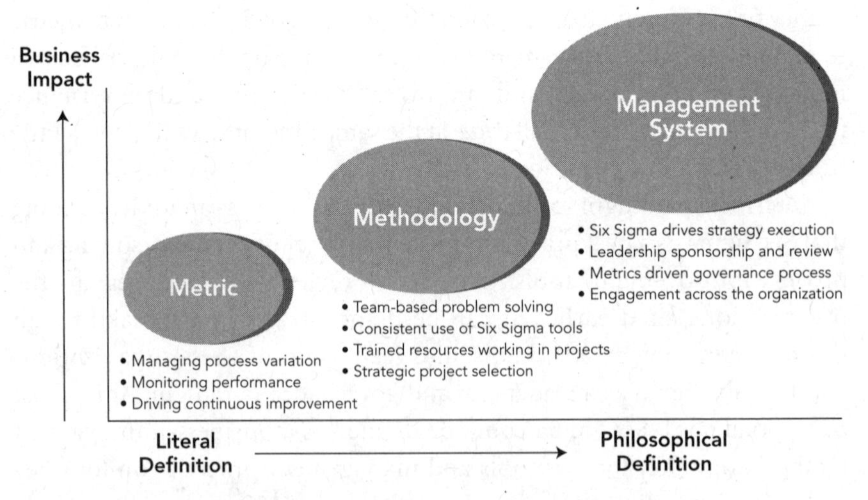

Figure 2-3 The Six Sigma maturity model.

In fact, in conversations with organizations and leaders that report dissatisfaction with the results of their Six Sigma efforts, we usually find sufficient knowledge and demonstration related to good use of metrics and methodology. All too often, however, we find teams practicing good methodology on low-level problems and process metrics that don't link to the overall strategy of the organization.

An early adopter of a leadership and governance framework built to both incorporate and support Six Sigma methodology was Motorola in the mid-1990s. The company described the framework as having the following activities that worked together to drive superior execution (Figure 2-4):

- Establish team alignment and awareness of the case for accelerated change through a baseline audit.
- Agree on and then vigorously communicate the winning strategy through the scorecard-development process.
- Agree on and then vigorously drive performance expectations through a vital few sets of operational metrics using the real-time dashboard approach.
- Agree on the critical processes and establish process owners.
- Drive organizational improvement through a rigorous Six Sigma governance process.

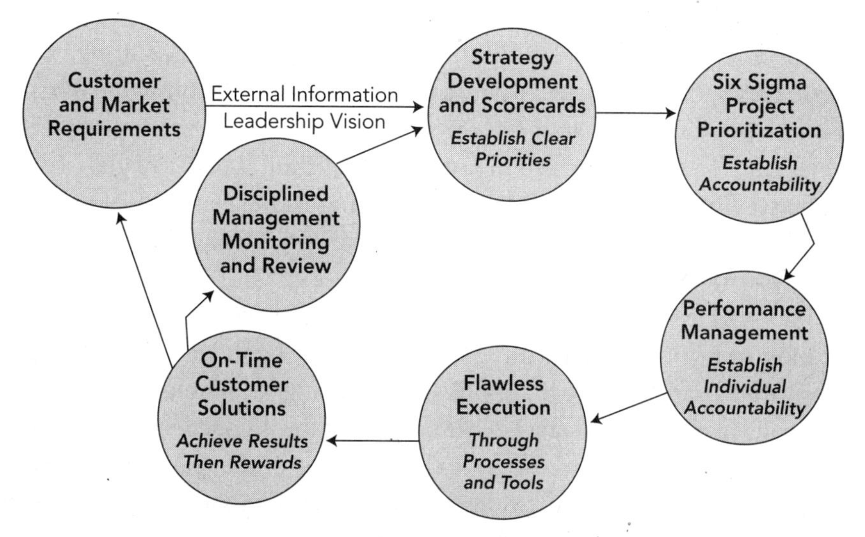

Figure 2-4 Six Sigma leverages key components.

- Drive adoption of Six Sigma leadership behaviors to enable and encourage front-line adoption of Six Sigma across the organization.
- Align individual performance-management processes with the Six Sigma management system.
- Build a reinforcing Six Sigma community of practice to drive organization-wide capability and continuous learning.

By following this framework in the mid-1990s and early 2000, Motorola was able to develop and market the highly successful Razor cellular phone while establishing itself as an early leader in sustainability practice.

The Collaborative Leadership and Governance Model

An example of an evolved leadership and governance model can be found in use at Jones Lang LaSalle, a global leader in real estate services and money management. The company, with approximately 30,000 employees worldwide, serves clients' real estate needs in more than 100 markets in over 35 countries on five continents.

Jones Lang LaSalle's full range of services includes agency leasing, property management, project and development services, valuations, capital markets, buying and selling properties, corporate finance, hotel advisory, space acquisition and disposition (tenant representation), facilities management (corporate property services), strategic consulting, and outsourcing. The company provides money management on a global basis for both public and private assets through LaSalle Investment Management. Jones Lang LaSalle has grown to become the leading supplier of global real estate services to corporate clients by building its business model around superior client relationship management and consistent delivery of high-quality worldwide service.

Corporate real estate organizations that operate globally are faced with a unique set of challenges with respect to aligning goals and implementing innovative practices across regions while driving local execution. This is especially true for Jones Lang LaSalle because of the unique relationships the company enjoys with top corporate clients. Many of the real estate services that Jones Lang LaSalle provides to corporate clients (such as Bank of

America, Motorola, Procter & Gamble, Xerox, and Microsoft) are performed at client sites and involve direct interactions with client employees and client work processes. Such services include providing engineering and maintenance support to client facilities, managing the physical movement of offices and whole businesses, and helping clients to reduce office and real estate space and associated costs.

Developing a Set of Leadership and Management Practices

Given this challenge, the strategic consulting team at Jones Lang LaSalle developed the *collaborative management model.* The model consists of a set of leadership and governance practices to guide client teams and their client partners in new ways of managing and governing business alliances within the shamrock organizational model that we described earlier. The model suggests that breakthrough performance occurs when all members of the leadership team have an equal seat at the leadership table. In the context of our shamrock organizational model, the leadership team would include representation from all four leafs—leaders from the core-employee team, leaders from key suppliers and strategic supplier alliances, representation from key subject-matter experts, and leaders from customer organizations. All members also need to have a common view and shared ownership of both the business outcomes and the operational performance required to achieve them. The model puts into operation a collaborative, team-based approach to strategy and planning, solution development, and shared decision making across the leadership team.

In these pages we will adapt the Jones Lang LaSalle collaborative leadership and governance model to demonstrate how the model can serve as a framework for driving sustainability initiatives across the four constituencies of a typical organization. Within the collaborative leadership and governance framework, the sustainability leadership team meets at quarterly intervals with "midpoint check-in" every six weeks. Unlike traditional management meetings, where suppliers and employees provide one-way reports and senior managers issue the next set of orders, the collaborative management model establishes an environment and an agenda focused on driving shared learning, innovative solutions, and breakthrough improvements to shared sustainability goals. This outcome is made possible through the following major elements of the model:

- A forward-looking continuous management process enabled through an established schedule of midpoint and quarterly leadership team meetings
- An established agenda focused on creating shared learning and innovative solutions
- A team commitment to a specific set of leadership behaviors
- A skilled external facilitator/coach capable of keeping the team focused on learning and innovation while practicing appropriate leadership behaviors
- A set of dashboards that deliver real-time performance data that enable fact-based decision making to drive timely execution of decisions and solutions

The collaborative management model is brought to life through a continuous cycle of dialogue sessions that follow an agenda designed to drive the leadership team toward creative strategy and solution discussions. The output of the session is an agreed-on set of strategies and actions that guide day-to-day resource allocation and management of execution activities.

All leaders have full responsibility for management and execution between leadership sessions. The leadership team is informed regarding progress through weekly updates to the operational dashboard.

On an annual and quarterly basis, the collaborative management team (CMT) focuses on listening to the voice of various constituents, prioritizing investment in key sustainability initiatives, and making adjustments to the overall sustainability strategy. On a monthly basis, the team focuses on driving operational priorities through careful review of sustainability metrics, project drill-downs, and key sustainability projects. On a weekly basis, the team is alerted to major variations in the sustainability metrics or in the status of critical projects. Meetings are held only when there is a need resolve a key performance alert, such as an unexplained spike in emissions or energy usage.

The system of workshops and infrastructure that drives this cadence is as follows.

Annual Planning and Strategy Development Workshop

The annual planning and strategy development workshop is focused on reviewing and understanding the organization's strategies. Through dialogue, with the support of an external facilitator, the team develops a shared understanding of the strategic inputs and implications to the business strategy. Once the strategic implications are understood, a series of facilitated activities drive the CMT to establish shared objectives (three-year horizon), key initiatives (this year's actions that support the objectives), performance metrics that indicate progress toward the objectives, and a set of Six Sigma improvement projects that focus on major cost-reduction opportunities and breakthrough process-improvement opportunities. Together the objectives, key initiatives, performance metrics, and Six Sigma projects become the CMT's strategic plan for the year. The plan is summarized in a one-page document that becomes the foundation for execution and review for the year.

CMT Strategic Planning Dialogue Agenda

- ▲ Create a shared understanding of business strategies.
- ▲ Conduct SWOT and performance-driver analyses.
- ▲ Agree on mission objectives and initiatives.
- ▲ Establish performance metrics.
- ▲ Launch Six Sigma improvement teams.
- ▲ Finalize the strategic plan.

Midpoint Dialogue Session

The monthly dialogue session is the primary vehicle for driving joint accountability for performance, integrated planning, shared understanding of client needs, and collaborative, innovative solutions. It is in these sessions that the linkage to key clients and resulting solutions is clearly established. Supported by structure and a facilitator/coach, client representatives present updates of client strategies and focus the team on emerging needs.

With these client needs as a backdrop, the team then reviews the performance status provided by the control center. The focus of this *operational review* is to develop a shared understanding of the sources of variation from operational plans. Where variation is apparent, process leaders will be on hand to walk through a process drill-down. Similarly, Six Sigma projects are reviewed for positive and negative variation to schedule and financial impact. Team leaders are on hand to review specific projects where CMT input is critical.

Once the operational reviews are completed, the team moves to dialogue focused on adjustments to the strategy and dialogue regarding emerging business-unit needs and innovative solutions to those needs. At a minimum, 50 percent of the focus of a monthly dialogue session is devoted to strategy and solutions. Each session focuses on developing shared understanding and enabling open and honest dialogue that drives innovative solutions and rapid execution.

Successful sessions result in leaders feeling confident that they can focus their attention on client relationships and strategy while the functional leaders focus on execution. These goals are achieved through a structured process and an established set of shared behavioral expectations supported by a skilled facilitator/coach familiar with both the strategy and the process.

Weekly Reviews and Alerts Dialogue

Critical to maintaining focus and driving execution is the timely resolution of unplanned variation in key performance metrics. In anticipation of occasional variations, a weekly check-in call is on the calendar of all CMT members. Participation in the call is determined by the nature of the issues to be discussed. The call is facilitated by the chief operations officer (COO) with support from the facilitator coach as required.

Collaborative Management Process Summary

When all the elements of the collaborative management process are knit together, a leadership team emerges that is focused on fulfilling customer needs with innovative solutions, driving execution, and trusting one another to contribute appropriately (Figure 2-5).

The Need for Real-Time Performance Data

The collaborative management model is informed through a set of *dashboards* (collectively referred to as the *guidance center*) that deliver real-

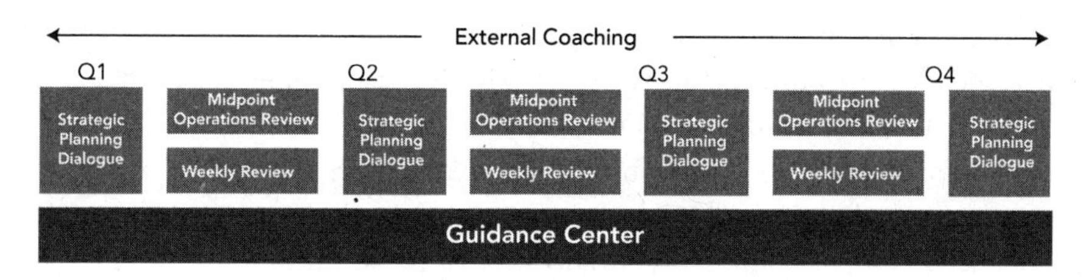

Figure 2-5 The cadence of workshops and infrastructure.

time performance data that enable fact-based decision making to drive timely execution of decisions and solutions. Whereas foundation of the CMT process is the strategic plan, the "nervous system" is the guidance center. This is an integrated set of management information displays that provide real-time or near-real-time updates and alarms related to critical performance objectives.

There are four displays:

- The *scorecard*, which tracks performance relative to strategic and financial impact
- The *operational dashboard*, which tracks performance relative to key operational performance metrics
- *Process drill-downs*, which enable leaders to click down into each operational process
- *Six Sigma projects updates*, which enable leaders to check status and financial impacts of key improvement projects

The guidance center enables members of the leadership team to spend less time collecting and interpreting data and more time communicating with clients and driving strategy.

The Role of the Facilitator/Coach

The role of the external facilitator/coach should not be underestimated in the collaborative management model. Leadership teams attempting to practice this model are learning a new set of processes, interactions, and personal behaviors. While driving superior execution, the team is operating in a collaborative, learning-focused dialogue.

Old behaviors and deep functional knowledge in certain areas often drive leaders to want to jump to what they see as a foregone conclusion informed by previous experience. Or they may be inclined to "trump" another team member's opinion because of either deeper expertise or higher organizational status. The result is a missed opportunity to develop a holistic system view and a potentially innovative solution.

Only a skilled external facilitator can manage the structured agenda while coaching team members on appropriate behaviors. There will be times when team members will feel that all the effort involved in attempting to collaborate is wasteful. The external facilitator/coach can push the team through these low spots.

Moving from Model to Practical Application

When all the elements of the collaborative management model are put into practice, leadership teams can reach new levels of sustained breakthrough performance, both operationally and financially. And when the model is extended to include suppliers and customers in the governance and decision-making processes, players win across the value chain.

It should be recognized that the collaborative management model is *aspirational* in nature. The team that developed it at Jones Lang LaSalle recognizes that complete implementation of the model, especially a model that drives deep collaboration between supplier partners and clients, requires mature relationships and deep trust across all parties.

The Key to Breakthrough Performance

Technical hurdles to delivering fully automated, real-time performance metrics must be overcome. And the tyranny of the urgent tends to derail the best-laid plans for proactive, forward-looking leadership dialogue.

However, as Jones Lang LaSalle introduces both the client-driven Six Sigma model and the collaborative management model to clients, more and more of them are recognizing that true breakthrough performance improvement and the associated financial gains can come only when a rigorous approach to problem solving and innovative solutions is combined with leaders who can facilitate collaborative governance and decision making across the value chain.

As a result, elements of the collaborative management model are in play across a number of key clients. Partnerships are being formed to drive collaborative Six Sigma projects. A number of business alliances are being governed through a common view of shared performance metrics. Governance boards are in place at a number of alliances where leadership is shared by the client and Jones Lang LaSalle. These developments point clearly to a future in which client-driven Six Sigma, enabled by the collaborative management model, will become the preferred management model for enlightened clients in need of best-in-class service delivery from professional service providers.

CASE STUDY

The Six Sigma Framework at Apex

Ron Brown is the general manager of a business unit of Apex. Ron recently replaced a general manager who had managed the business for 10 years, and he found himself facing a number of challenges.

With a solid reputation as financial wizards, Apex and its employees built their success on a formula of one new financial product breakthrough after another. As a result, the company sustained a 20 percent growth rate for the past 10 years.

After a decade of growth, however, Apex found that its customers had lost their appetite for investment in speculative finance. Instead, they demanded more performance at a lower price and more responsive service on currently installed products and systems. New competitors found ways to copy Apex products and deliver them at a lower price. As customer order rates declined dramatically, Apex found it difficult to reduce transactions costs or attract new customers for its higher-margin products. Cash flow was suffering, and Apex could not fund new-product development opportunities as it had in the past. And even though Ron had made very public statements to shareholders, employees, and customers about the Apex commitment to building a sustainable product platform and improving brand loyalty and profitability through sustainability initiatives, the initiatives received no

CASE STUDY

traction. They were lost in the day-to-day struggle to revive slumping product sales. Needless to say, Ron was not sleeping well.

The leadership team at Apex had deployed Six Sigma improvement teams to drive specific process improvements and energy cost-reduction initiatives for the past two years. While each member could point to specific projects that had shown some solid improvements, it didn't seem that those improvements were creating measurable financial gain. Margins continued to decline. The organization was dangerously close to negative cash flows, and energy consumption and costs continued to increase.

Reflecting on Six Sigma project-level wins as well as disappointments, Ron was convinced that this was not the time to drop Six Sigma and try a different approach to improvement. He believed that there was a way to build on the experiences, positive and negative, of the past two years and develop a more comprehensive, bottom-line-oriented approach to Six Sigma business improvement.

Ron pulled his leadership team together for a working session focused on a key question: "How can we achieve more consistent weekly results *as well as* more sustainable improved performance while implementing our sustainability initiatives?"

Starting with this question, members of the leadership team discussed their performance over the past two years. They agreed that Six Sigma had been a useful approach to solving some critical problems. Order rates were improving. Product launches were going more smoothly. Six Sigma teams had reduced energy consumption in their data centers dramatically. Numerous processes were operating more efficiently and more productively. At the same time, the team was perplexed by the fact that overall organizational performance, as measured by profitability, shareholder value, and carbon footprint, had not seemed to improve despite these efforts.

It seemed to Ron that the team had achieved reasonably good adoption of Six Sigma tools for support in discrete problem-solving situations. On just about every problem, Ron was seeing much deeper levels of analysis and much better identification of root causes before

CASE STUDY

solutions were being implemented or investments were being made. In addition, during the past two years, 20 black belts and 50 green belts had been certified with projects that demonstrated an average savings or incremental revenue gain of $200,000 per project. Still, Ron said, overall organizational performance remained flat or was declining in certain key areas (Figure 2-6).

Working	Not Working
• Improved yields/reduced scrap in production	• Other functions not fully on board
• Certified black belts and green belts access the organization	• No documented return on investment dollars spent on training
• Project level savings in exess of $5,000,000	• Savings don't seem to show in bottom line
• Problem solving tools helping teams get to root causes	• Leaders are not supporting breakthrough investments
• Many enthusiastic supporters	• Employees wonder if the commitment to change will last

Figure 2-6 Effectiveness of Six Sigma at Apex.

So Ron brought together his leadership team for a second Six Sigma strategy discussion. He shared the client-driven Six Sigma model with the team and emphasized the importance of its five components operating as an integrating system (Figure 2-7).

Five Focus Areas of Client-Driven Six Sigma:

I. Voice of the Customer
II. Performance Metrics
III. Lean Process Design
IV. DMAIC Process Improvement
V. Governance

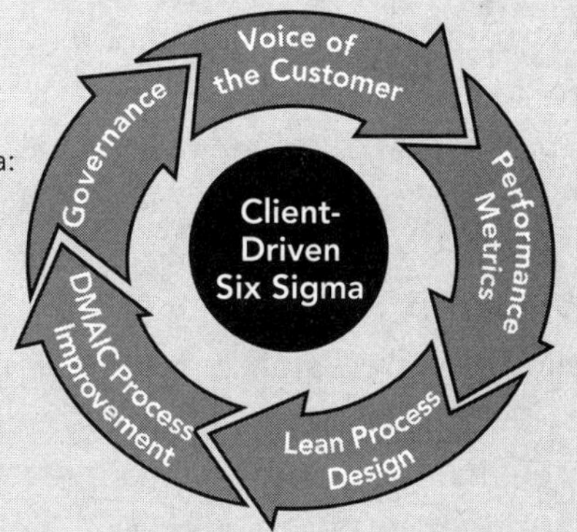

Figure 2-7 Client-driven Six Sigma model.

CASE STUDY

The team realized that, intuitively, they all understood the principles of the model. Customer requirements should drive the metrics that guide the business, and those metrics should absolutely determine which processes were critical to the success of the business. Critical investments and resource decisions should be made in ways that enabled the core processes to deliver maximum value to customers.

At the same time, however, the team admitted that, in practice, this was not the way they ran the business. Each leadership team member operated specific areas of responsibility as if they were stand-alone businesses. Accountabilities and rewards were established primarily against financial objectives. There had always been an unstated pact between team members: "You stay out of my business, and I'll stay out of yours. If we all perform well within our function, the business will prosper."

Aligning Apex with the Client-Driven Six Sigma Model

At this point Ron asked the team to describe ways that they could more closely align Apex with the client-driven Six Sigma model.

Mary, the vice president responsible for sales and marketing, had good data that showed a clear set of customer priorities. Henry, the chief innovation officer, had already been in discussions with Fred, the vice president of supply-chain operations (procurement, production, and order fulfillment), about how they could work more closely together to get products to market more rapidly. Jack, the chief financial officer, could see that while financial metrics represented the "bottom line" and financial discipline was critical, financial metrics actually were after-the-fact measurements. He had always been frustrated by the team's inability to predict changes in financial outcomes in time to make operational adjustments. Jack believed that operational metrics, aligned with customer requirements and linked closely with the key operating processes, would allow the team to run the business with a more hands-on, proactive approach rather than an end-of-the-month, looking-back perspective. With all this in mind, the team developed a client-driven Six Sigma framework for Apex (Figure 2-8).

CASE STUDY

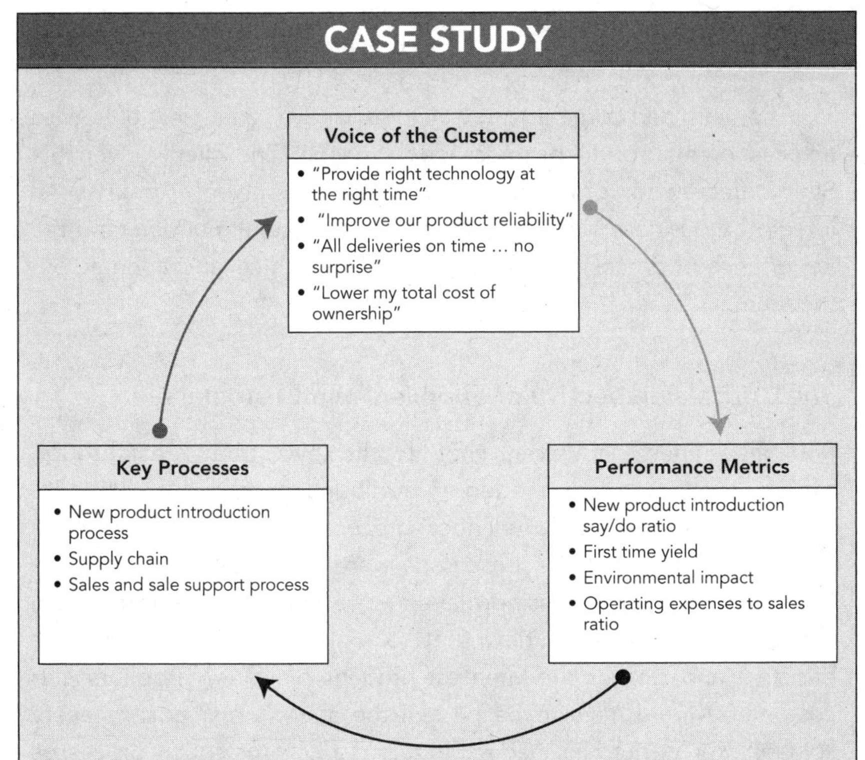

Figure 2-8 The client-driven Six Sigma framework for Apex.

It was clear to the team that Apex could be run in a more integrated fashion. The obvious challenge was putting the framework into day-to-day practice. Breakthrough performance requires a team that is willing and able to lead in a very hands-on and collaborative way. Superior execution occurs when a team has full operational awareness and is able to drive activity across the entire business.

The new client-driven Six Sigma framework for Apex would enable the leadership team to see opportunities more holistically and then focus cross-functional teams on high-leverage points of improvement. Able to see the entire business system, and no longer concerned with stepping onto each other's turf, the team could sponsor and lead Six Sigma projects that truly could search out and eliminate the root causes of their toughest business-performance problems. Driving solutions to true systemic root causes would be the key to translating Six Sigma improvement to sustainable financial impact.

CASE STUDY

The team had come a long way in expanding their view of how to achieve breakthrough business performance. The client-driven Six Sigma framework provided an enlightened view of Apex as an integrated business system, and it helped the team to understand the importance of process ownership combined with collaboration across the team.

The Critical Perspective of Suppliers and Customers

Ron also knew, however, that to achieve true breakthrough performance, the team needed an additional perspective. When he looked closely at each of the critical processes, it was clear that some key roles in each of the processes were not represented on the leadership team. The missing players were key suppliers and certain critical customers. Ron knew that to make Apex processes most effective and deliver the greatest amount of value, suppliers and customers would have to be part of the analysis and part of every solution. Ron wanted his leadership team to view the critical processes beyond the Apex boundaries—from suppliers at one end to customers at the other (Figure 2-9).

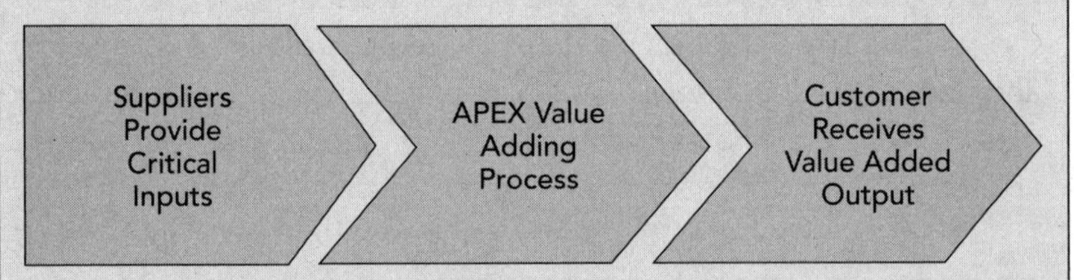

Figure 2-9 The Apex value chain.

Ron realized that the total-value-chain concept represents another idea that is intuitively obvious but has significant implications for how his leadership team would operate moving forward. The broad implication was that the team was going to have to find mechanisms for bringing customers and suppliers into the Apex leadership and decision-making process. This would go beyond traditional notions of supplier

CASE STUDY

councils and customer advocacy boards. Customers and suppliers would have to participate directly in the Six Sigma improvement projects. Beyond this, though, Ron knew that certain suppliers and key customers would have to become part of the leadership and governance of the Apex integrated business-process model.

The Need for Discipline and a Collaborative Leadership Environment

At the same time that leadership would be challenged with the need to collaborate much more closely with customers and suppliers, leaders also would be challenged with the need to be more disciplined and rigorous as a management team while simultaneously being much more collaborative in their analysis and decision making (Figure 2-10).

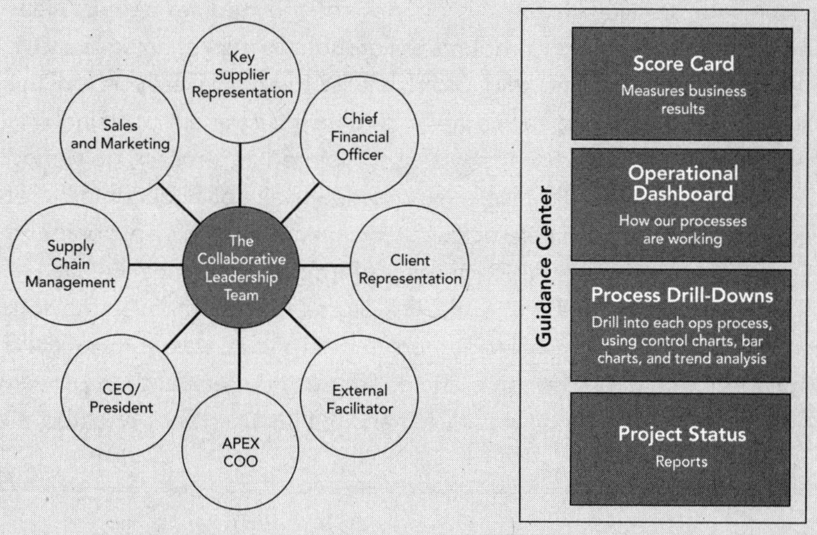

Figure 2-10 The collaborative management model.

Breakthrough performance requires a team that is willing and able to lead in a very hands-on and collaborative way. Superior execution occurs when a team has full operational awareness and is able to drive

CASE STUDY

activity across the entire business. So the team was introduced to the collaborative management model as a tool for integrating customers and suppliers into their leadership and governance model. The team invited key customers and suppliers to join them in a two-governance meeting that would occur on a quarterly basis.

Although both customers and suppliers were reluctant at first to openly share ideas and concerns, the participants learned that Apex was serious about open and honest dialogue and that it was serious about implementing those ideas. The entire team learned that they shared common values and concerns relating to the environment and their collective impact on the environment. Dialogue about environmental impacts soon led to a sustainability initiative that stretched across the entire value chain.

In this collaborative leadership environment, work flows seamlessly across the value chain. Expectations established at one end of the value chain are delivered on, according to the requirements. New, environmentally friendly products are brought to market on time and at the forecasted volume and profit levels. Everyone delivers on the promises they make to each other and to their customers. Sustainability projects are resourced properly and completed as forecast, delivering the financial impacts that were expected. Sustainability becomes a set of initiatives that are linked across the value chain. In this environment, sustainable, breakthrough business performance becomes a reality.

By combining the use of the Six Sigma management framework with the collaborative management model, Apex was able to build collaboration across its value chain. The results were both improved financial performance and a sustainability initiative that benefited all parties.

Chapter Summary—Key Points

- If leaders honestly buy into the business case for sustainability, then they understand that having a solid commitment to sustainability supported by a robust operational sustainability plan is a critical factor in the long-term success of their organization.
- Every improvement project requires careful attention to the leadership and change-management implications associated with the proposed improvement.
- Pyramid-like organization models (leader at the top, employees at the bottom) are not flexible enough to adapt to the rapid pace of change in external markets that most organizations face. And the command-and-control style of pyramid-shaped organizations is no longer attractive to workers interested in flexibility, diversity, and autonomy.
- Today, people increasingly work
  - In networks and alliances
  - In collaborations
  - On projects and initiatives of varying duration
  - By choice, according to the circumstances of the person
- The flexible organizational response to these factors best suited for deploying a Six Sigma sustainability program replaces the pyramid with a shamrock with these four leafs:
  - Core workers
  - Strategic supplier alliances
  - Independent subject-matter experts
  - Customers
- Good metrics and process-management methods are not sufficient for organizations that desire breakthrough improvements over time. Successful business champions view Six Sigma as an integrated set of leadership and governance practices supported by disciplined use of specific behaviors.
- The collaborative management model consists of a set of leadership and governance practices to guide teams in new ways of managing and governing business. The collaborative management model is made possible by:
  - A forward-looking continuous management process enabled through an established schedule of midpoint and quarterly leadership team meetings

- An established agenda focused on creating shared learning and innovative solutions
- A team commitment to a specific set of leadership behaviors
- A skilled external facilitator/coach capable of keeping the team focused on learning and innovation while practicing appropriate leadership behaviors
- A set of dashboards that deliver real-time performance data that enable fact-based decision making to drive timely execution of decisions and solutions

CHAPTER 3

The Sustainability Transfer Function

We've found over time that there are two schools of thought when it comes to *performance excellence.* One approach is process-based, and the other is cause-effect-based. Each approach has in mind the same endpoint: Define the critical few activities that, when managed, will create results consistent with customer and shareholder expectations. This set of results-creating activities comprises the *transfer function.* The intention of this chapter is to explain how to develop the transfer function for your sustainability program and to show relevant examples that could be used as your starting point. By applying the principles described in this chapter, you should be able to create a transfer function as a management tool for *sustainability excellence.*

Transfer Functions and Why They Are Important

The process-based approach was conceived by Motorola and General Electric, where a company's operations are composed of core processes, such as product development and sales, and all activities flow down from this set of core processes in a neatly organized series of process levels. This approach is commonly referred to as *process excellence.* If level 1 consists of your core process of "acquire customers," for example, then an example of a level 2 process might be "qualify customer leads," and a level 3 process within that scope might be "determine customer's ability to pay."

The other common approach to performance excellence sees business success as a cascading series of causes and effects. The entire business could be mapped as one big fishbone diagram where each driver is instrumented and measured. This approach was developed on the assumption that, in many instances, it is more trouble than it's worth to try and fit every important business activity together in a linear series of supplying, processing, and outputting.

Both process excellence and business-performance excellence make use of the transfer-function structure in organizing a company's improvement efforts. The transfer function is an important tool for quantifying business performance and for identifying improvements by understanding the critical few drivers for success. At the top of the transfer function is the "big *Y*," the high-level business outcome that shareholders and stakeholders want to achieve (think revenue). Big *Y*'s cannot be controlled directly. For example, if your big *Y* is "revenue," your next steps are to identify the drivers of revenue and to keep breaking down the drivers until you get to the work you can control directly. (It would be great if you could control revenue directly—then all your shareholders would have to say is, "Go get us more revenue," and Shazam! You'd do it!) Alas, the primary objective (big *Y*) "revenue" cannot be controlled directly; therefore, to create it, we must understand and activate its drivers. The same is true for sustainability.

If the overall result is the big *Y* and it is not directly controllable, what do we call the actions we can control? We refer to them as *X*'s. The nomenclature is rooted in basic algebra; that is, *Y* is a function of *X*. Or *Y* is a function of multiple *X*'s: *X*1, *X*2, *X*3, . . . , *Xx*. Big *Y*'s also include little *y*'s, or intermediate (but still not directly controllable) results. And little y's may include big *X*'s, which are generally unactionable (but descriptive) categories of activity. The final layer of the transfer function is the little *x*, the actions one can take to create a result. Thus we have

Results: Big *Y* is a function of little *y*'s; for example, revenue (big *Y*) is a function of sales leads generated (little *y*).

Descriptions: Little *y*'s are (sometimes) categorized by big *X*'s; for example, sales leads (little *y*) are categorized by geographic region (big *X*).

Actions: Big *X*'s are a function of little *x*'s; for example, leads in each geographic region (big *X*) are a function of the number of sales calls and other direct actions (little *x*'s).

In this chapter and in this book in general, we will refer to *Y*'s and *X*'s as shorthand for results and actions. As you go through this book, you might be tempted to question whether a particular *X* is really a *Y* or vice versa. The only important item for you to bear in mind is that results have causes and that to take action you must identify the drivers of the result you seek. In the overall cause-effect chain, a result in one part of the chain may look like a cause to some other part of the chain. This is okay.

It is also important to keep in mind that not all actions have equal impact on the result. Seek out the critical few drivers of the desired result. Whether this is done using statistical tools, such as a regression analysis, or simply through logic and experience, focus should be placed on the *critical few* drivers, or *x*'s. You will run across situations where the interaction between otherwise minor factors is important (which is one of the reasons why sustainability is considered a systems challenge). In general, though, start with the *x*'s that are the biggest needle movers.

Identifying the transfer function for your company's sustainability efforts should improve the efficiency and effectiveness of planning, measurement, and improvement. Your managers and staff should feel that there is logic to your company's approach and should be able to identify their individual roles in the program.

Building the Sustainability Transfer Function

A commonly referenced definition of *sustainability* (Figure 3-1) comes from Brundtland[1]:

> Sustainable development is development that meets the needs of the present without compromising the ability of future generations to meet their own needs. It contains within it two key concepts: the concept of needs, in particular, the essential needs of the world's poor to which overriding priority should be given; and the idea of limitations imposed by the state of technology and social organization on the environment's ability to meet present and future needs.

A diagram of the Brundtland definition is shown in Figure 3-1.

Technology and social organizations take form as institutions of one type or another. These institutions—businesses, governments, communities, charities/associations, and schools—tend to consider there to be three types of sustainability: economic, environmental, and social. This is referred to as a *triple bottom line* for an organization that focuses on profit, planet, and

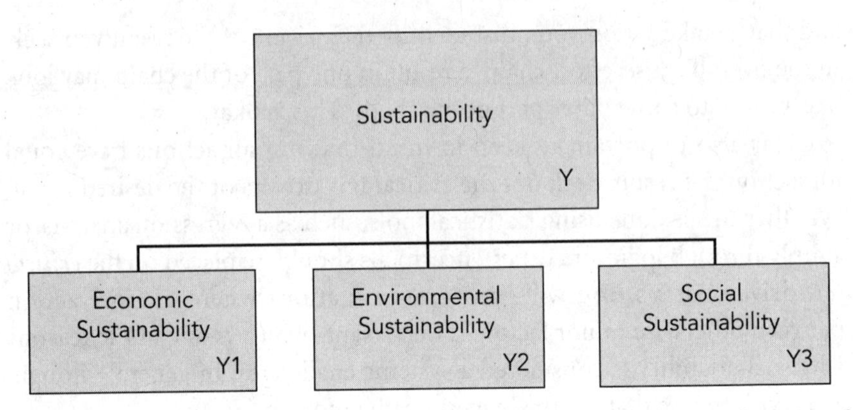

Figure 3-1 The top-level transfer function: Overall sustainability as a function of economic, environmental, and social sustainability.

people. Each category overlaps each other category in important ways: Economic growth may be attained at the expense of the environment (e.g., deforestation for lumber) or social good (e.g., child labor). Environmental phenomena may disrupt commerce (e.g., a volcano interrupting air travel). Or social factors such as regulation may encroach on the rights of private capital. Social media can create economic opportunity, and there is certainly an abundance of social media devoted to environmental issues. We observe that the world we live in seems less sustainable when developments in these three areas become out of balance. When incongruities occur with frequency and/or amplitude, one part of the system will complain. And because the tools for maintaining a sustainable environment are largely social, one could consider that current sustainability efforts are aimed at bringing economic concerns further into the balance. Depending on your political views, you may feel that industry is being asked to do more or less than it should do.

Since there is no shortage of books on the topic of economic sustainability, the focus of this book is on addressing environmental sustainability within the economic (corporate) context. We are doing this by using an inherently business-oriented framework, Six Sigma, as the organizing structure for environmental sustainability. The desired result is to show how economic and environmental concerns can be resolved simultaneously.

Because with regard to natural resources the planet is a closed system, we must examine our use of natural resources to live up to the aspirations of the Brundtland definition of sustainability. One obvious constraint is that in a world of finite resources, even taking into account increasing levels of productivity, we cannot use more resources than can be replenished in time for future generations to meet their needs. Using more resources than can be replenished in time for future generations of people is inherently unsustainable. When considering resource use patterns in various parts of the world, increasing productivity and the negative impact of waste, and increasing planetary population, it is easy to get overwhelmed and to assume that someone else smarter than we are must be solving this huge system problem. Unfortunately, aligning economic incentives to do so and making industrial systems sufficiently sensitive to nuanced feedback from environmental systems are crushingly difficult. We need rules to govern our stewardship of resource quantity and quality. And we need to lace these rules into our business processes.

The next level in our transfer function for environmental sustainability takes these rules into account, thanks to a framework called the *Natural Step*. The Natural Step is an internationally recognized science-based framework for achieving sustainability in concrete ways. Using the Natural Step helps us to address such questions as

- What are the basic problems that make our global society unsustainable?
- How are we contributing to these problems?
- What can we do today and tomorrow to stop contributing to these problems?

According to the Natural Step, there are three critical rules that a system such as a business must not break in order to be environmentally sustainable and one rule for social sustainability. According to the Natural Step, in a sustainable society, nature is not subject to systematically increasing:

1. Concentrations of substances extracted from the Earth's crust
2. Concentrations of substances produced by society (concentration could be physical or gaseous materials)
3. Degradation by physical means

And, in that society,

1. People are not subject to conditions that systematically undermine their capacity to meet their needs.

By adding these results to the sustainability transfer function (Figure 3-2) as level 3, you can see how the cause-effect framework grows and

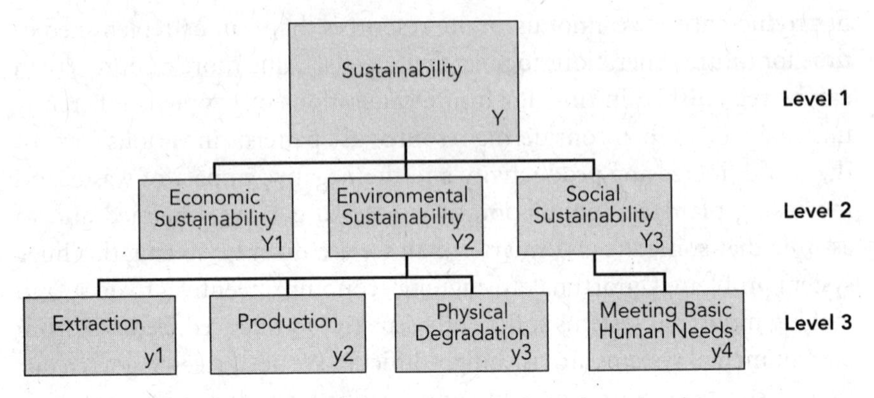

Figure 3-2 Sustainability transfer function to level 3.

becomes more specific. For notation purposes, we also have shifted from big *Y*'s (e.g., Environmental Sustainability *Y*2) to small *y*'s (e.g., Extraction *y*1). This notation is somewhat a judgment call, but we feel that it fairly represents the shift from a concept-level result with complex composition to a still complex but more obvious result.

One thing you might notice that this internationally recognized, science-based framework does not say is, "Don't emit carbon dioxide into the atmosphere." Depending on the source of the carbon, carbon emissions can be considered a no-no under systems condition 1: Don't subject nature to systematically increasing concentrations of substances extracted from the Earth's crust. Carbon emissions are generated when fossil fuels, which are extracted from the Earth's crust, are burned. When the power plant burns the coal to create energy, carbon dioxide is emitted into the atmosphere.

The reason why carbon footprint gets so much attention is its role in a subset of environmental sustainability: climate change. Carbon emissions build up in the atmosphere along with other greenhouse gases, and the insulating properties of these gases cause an aggregate warming effect on the Earth. (Carbon dioxide isn't the only greenhouse gas or even the most potent one from a climate-change perspective.) Of all greenhouse gases, six are addressed by the Kyoto Protocol and are referred to as the *Kyoto gases*: carbon dioxide, methane, nitrous oxide, hydrofluorocarbons, perfluorocarbons, and sulfur hexafluoride.

In our overall transfer function, carbon footprint is a direct result of burning fossil fuels. The World Resources Institute published the defining framework for reporting carbon footprint called, *The Greenhouse Gas Protocol.* According to this publication, there are three buckets of carbon emissions and one for noncarbon emissions:

1. *Scope 1:* Generation of electricity, heat, or steam; manufacture or processing of chemicals (e.g., cement, aluminum, ammonia); combustion of fuels in company-owned vehicles; fugitive emissions from leaks
2. *Scope 2:* Electricity purchased for own use
3. *Scope 3:* Other carbon emissions
4. Other noncarbon emissions

According to *The Greenhouse Gas Protocol*, companies are responsible for reporting carbon emissions from scope 1 and scope 2 within their operational boundaries. The company also may report relevant emissions from scope 3 sources. Chlorofluorocarbons are not part of the Kyoto Protocol but may be reported separately, as may be carbon emissions from the burning of biofuel.

Up to now, we have provided you with the transfer-function content based on international management and reporting frameworks that have studied issues such as carbon emissions for years. To make this transfer function work for your company, you will have to use your own analysis of which results (*y*'s) are relevant in your business and which drivers (*x*'s) contribute to those results in your company. For example, although reporting scope 3 carbon emissions from employee business travel is optional, Cisco has determined that it is important to its business and therefore is something Cisco chooses to manage and reduce. According to the U.S. Energy Information Agency,[2] about 34 percent of energy-related greenhouse gas emissions is from the transportation sector. This supports the case for reducing transportation, which Cisco addressed through a variety of initiatives, resulting in a 39 percent reduction in air-travel greenhouse gas emissions during the period fiscal year (FY) 2006 to FY2009[3] (Figure 3-3).

The Transfer Function for Office Waste

Waste-reduction targets are a very common component of corporate environmental sustainability programs. According to the U.S. Environmental Protection Agency (USEPA) in 2006, U.S. residents, businesses, and

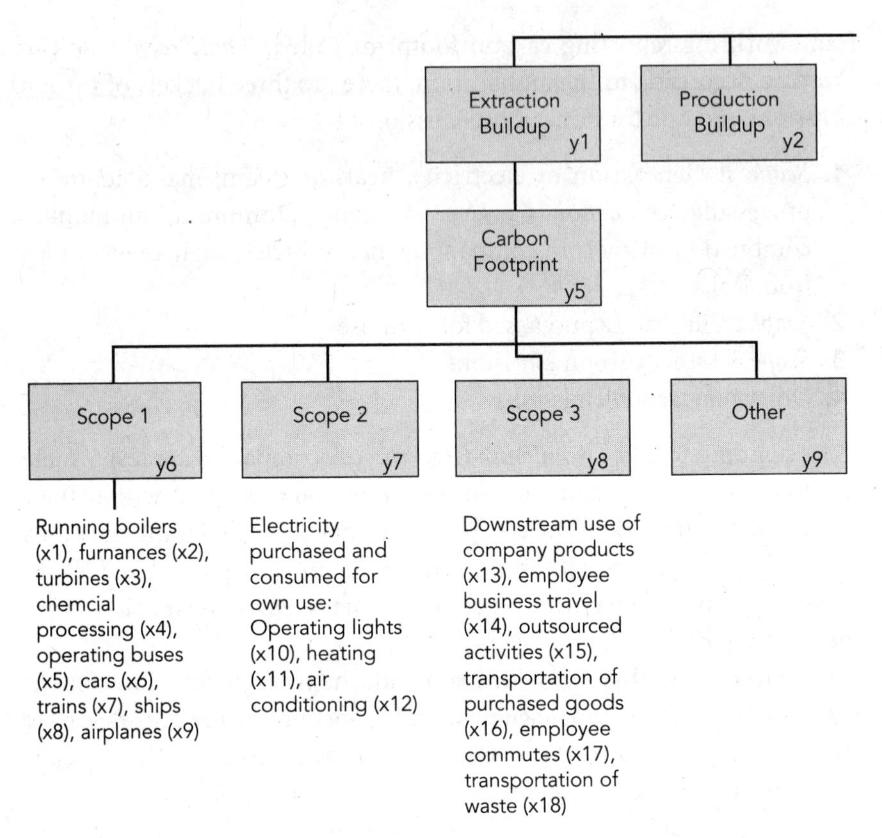

Figure 3-3 The sustainability transfer function to level 4: carbon footprint.

institutions produced more than 251 million tons of municipal solid waste,[4] which is approximately 4.6 pounds of waste per person per day in the United States. Municipal solid waste goes to landfills. Materials decomposing in landfills create gases that are sometimes burned for energy or otherwise escape to the atmosphere and add to the greenhouse effect. Landfills also contribute to the systematic degradation of the physical environment.

Part of the municipal solid-waste stream is paper. According to the Regional Environmental Center for Central and Eastern Europe, about 650,000 tons of paper are produced daily worldwide, whereas 500,000 tons are thrown away because they are not reintroduced into the paper production cycle. Solid waste and the paper-based portion of that waste are important environmental sustainability issues.

Corporations manage business processes and operate facilities that generate a variety of waste streams. Some examples of corporation-generated waste streams from nonmanufacturing processes are

- Mixed paper (which will be used for a more detailed transfer-function example later in this chapter)
- Cardboard
- Glass or plastic bottles (beverage containers)
- Cans
- Plastic
- Foam
- Landscape waste
- Fluorescent lights
- Toner
- Furniture
- Food (e.g., from cafeterias)
- Scrap metal

Although the use of office paper may not be the largest purely environmental sustainability issue in your business or even your largest waste-stream management issue, it is relatively easy to understand and is always an area of interest on the part of office employees. Therefore, it makes a good example for further defining the sustainability transfer function down to an actionable level. The office paper process is relatively straightforward (Figure 3-4).

Even this simple process, however, introduces a number of important considerations for corporate environmental sustainability planning and management:

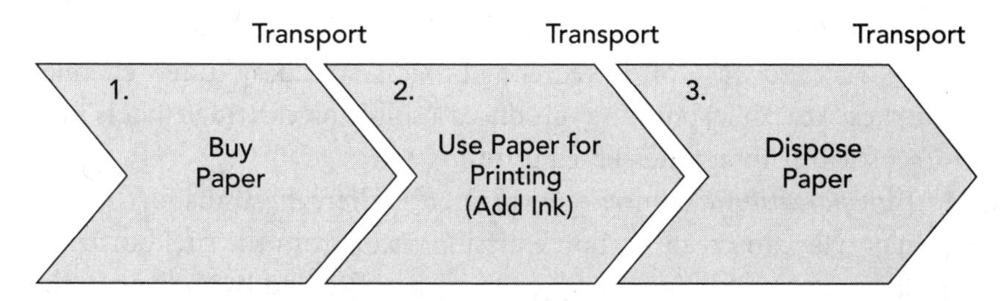

Figure 3-4 The office paper process.

1. *Tonnage of paper waste created and landfill diversion rates.* When companies and their employees start recycling paper waste, one of the common metrics used is diversion from landfill (i.e., the percentage of waste that goes to recycling compared with the overall amount of waste generated). It is important to consider that as a company takes an end-to-end approach, the total amount of materials used can be reduced. Therefore, landfill diversion rates should increase initially; then, as the company reduces its paper use and the amount of paper it procures, the proportion of diverted paper will plateau and then decrease. It is important to measure total waste in addition to diversion rates.
2. *Procurement practices such as recycled content requirements.* From an environmental sustainability standpoint, purchasers of office supply paper should consider the proportion of postconsumer waste content in the paper itself. It is not difficult to find paper with 30, 50, or even 100 percent postconsumer fibers. Some procurement departments may adopt standards for paper that is warranted to come from sources certified as sustainable by the Forest Stewardship Council or the Sustainable Forestry Initiative (see below for more information about the upstream impact of demand for office paper).
3. *Product-use behaviors and product-replacement technologies.* When companies start to consider why they need paper to begin with, they also can consider whether the paper is being used efficiently and whether there are nonpaper means for accomplishing the same goals. Do employees print double-sided? Do employees print more copies/drafts than they need? Can digital technologies be used instead of paper to fulfill the same communication or documentation needs? Even the choice of fonts and inks can have an impact on environmental issues. The University of Wisconsin in Green Bay changed the default font[5] in its e-mail system from Arial to Century Gothic because it uses about 30 percent less ink to print the same information. Soy-based inks are easier to wash off in recycling processes, leading to few chemical issues; Xerox Corporation[6] produces a solid-ink cartridge that is more ecofriendly than toner-based printing.
4. *Transportation distance.* As explained earlier, the transportation of materials can create carbon emissions (i.e., from the fuel burned by the trucks that bring the paper to your office and by the trucks that cart away waste). Reducing paper procurement and solid-waste pickup

will reduce transportation-related carbon emissions as well as operating costs.

The scope of the transfer function is broadened considerably when an end-to-end lifecycle approach is used. Figure 3-5 shows a more complete picture of the paper-related drivers of sustainability.

Because forests play an important role in the environmental system, much attention has been paid to the forest-management practices of the commercial wood and paper industry supply chain. Different stakeholder groups have defined forest-management sustainability in different ways, with the essential differences coming down to the relative emphasis on economic, environmental, and social issues. The life cycle of paper products constitutes an excellent example of how difficult it is to separate the three sustainability big *Y*'s in our transfer function. We must consider the human judgment that determines how each company approaches its responsibilities in this area. These management decisions become represented on the transfer function based on stakeholder and shareholder influence.

The environmental sustainability of step 1 in Figure 3-5, "Harvest wood from forest," is a function not only of the effectiveness (even fuel efficiency) of harvest machinery but also of forest-management practices of the landowner. Sustainable forest-management considerations include several interrelated factors:

1. The quantity and replacement cycles of trees
2. The health of the trees
3. The ability of the forest to support various native species
4. Soil protection from erosion
5. Maintaining the integrity of underground water supplies
6. The ability of trees to store carbon dioxide

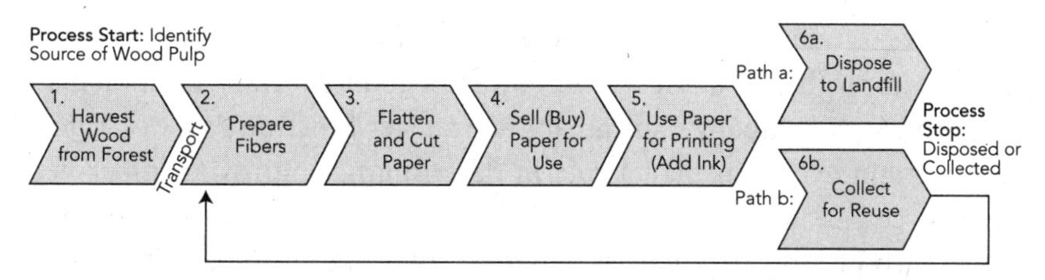

Figure 3-5 The paper production level 1 process (lifecycle approach).

The environmental sustainability of step 2 in Figure 3-5, "Prepare fibers," is a function of water use, chemical use for the purpose of bleaching, and other high-energy-consuming equipment. Step 3, "Flatten and cut paper," is also very energy-intensive. Public consideration of the issues and actions along each step in the paper-production process has led to a number of certifications that may be sought to show a commitment to environmental stewardship. Some examples are as follows:

- *Forest Stewardship Council certification*—focus on forest management and related social issues.
- *Green Seal certification*—ensuring postconsumer content of paper products.
- *Processed chlorine-free*—manufacturing paper without chlorine.
- *Green-e certified*—certifies the renewable-energy credits purchased by some paper-production companies.

Because most companies are only paper users, not papermakers, their ability to affect change across the paper-production life cycle is limited to procurement, use, and recycling practices. Designing your transfer function for what you can control, informed by the actions and issues upstream and downstream in the supply chain, is important. From the perspective of the environmental sustainability of the paper-production life cycle, the critical few *x*'s on your company's transfer function may reside in your procurement process (Figure 3-6; see also Chapter 9 for supply-chain management considerations).

The Role of Location

To this point, we have not integrated into our transfer function any factors related to location. As mentioned early in this chapter, location, as a descriptive category, is not actionable but is notated as a big X because it's generally considered a step in between a result (Y) and an action (x). This is not to imply that location is unimportant. On the contrary, location plays a key role in a company's sustainability transfer function as a descriptive variable and even can play a role in determining improvement-project priority.

In carbon-footprint analysis, location is often a shortcut to determining the carbon factor for electricity consumption in scope 2 carbon emissions

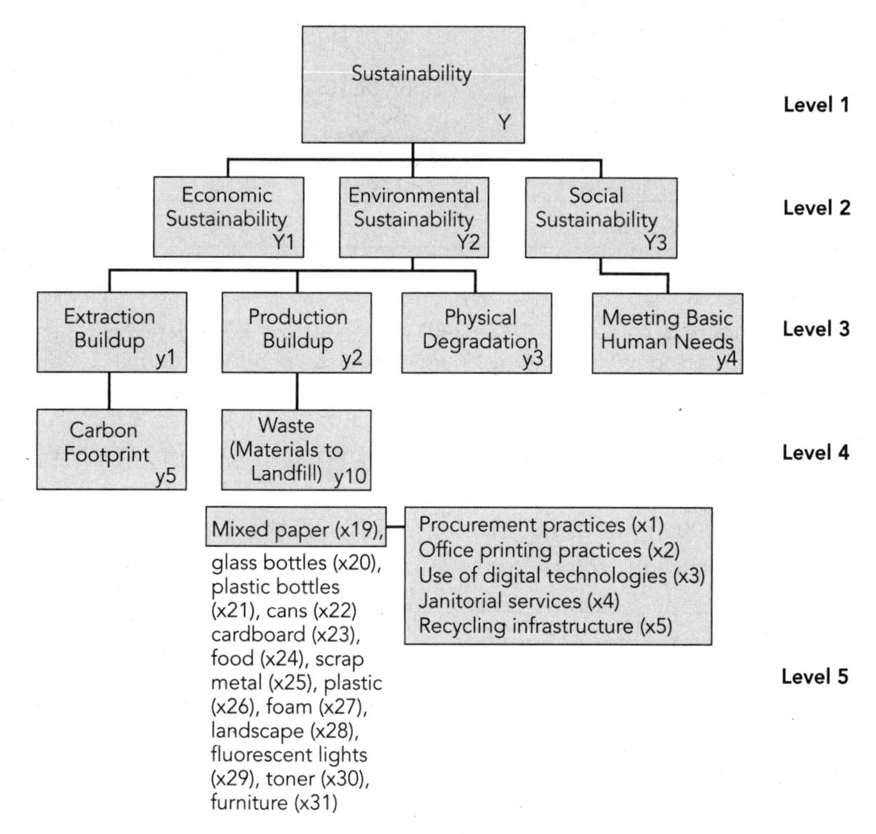

Figure 3-6 The sustainability transfer function showing waste (*y*10), the contribution of paper (*X*19), and the drivers of office paper environmental sustainability (*x*1–*x*5).

inventory. This is to say that electricity in different regions of the world and in different parts of the United States is generated with different carbon loads depending on the source of the energy. If electricity is generated from hydropower sources, it is relatively clean and has a low carbon factor. If electricity is generated from coal, it has a higher carbon factor. The carbon footprint for scope 2 emissions is calculated by multiplying the quantity of electricity consumed by the carbon factor for the source of that electricity. Consumption is determined from utility bills, and the carbon factor can be looked up in the USEPA eGRID table. Figure 3-7 provides examples of carbon factors by eGRID subregion using eGRID data published in 2004 for parts of Alaska (77 percent nonrenewable energy sources and 23 percent renewable energy sources), Arizona (94 and 6 percent), and California (68

eGRID Subregion Acronym	eGRID Subregion Name	Annual Output Emmission Rates		
		Carbon Dioxide (CO_2) (lb/MWh)	Methane (CH_4) (lb/GWh)	Nitrous Oxide (N_2O) (lb/GWh)
AKGD	ASCC Alaska Grid	1,232.36	25.60	6.51
AZNM	WECC Southwest	1,311.05	17.45	17.94
CAMX	WECC California	724.12	30.24	8.08

Figure 3-7 Sample EPA eGRID carbon factors (2004 data). (www.epa.gov/cleanenergy/energy-resources/egrid/index.html)

and 32 percent). Applying the carbon factors from this chart for 1,000 MWh of electricity would yield scope 2 carbon emissions of 1,232,000 pounds in Alaska, 1,311,000 pounds in Arizona, and 724,000 pounds in California. Applying the carbon factors from this chart for 1,000 MWh of electricity would yield scope 2 carbon emissions of 1,232 pounds in Alaska, 1,311 pounds in Arizona, and 724 pounds in California.

When we structure a transfer function to include categories such as location, we use the big X notation for each category (Figure 3-8). When reporting scope 2 carbon emissions data under this structure, we would organize our data by state (or by eGRID zone, each of which has a slightly different boundary than a state). You can reduce your company's scope 2 carbon emissions by moving operations from a state with a higher carbon factor to one with a lower carbon factor (e.g., from Arizona to California in this example). In this way, you might consider geography to be an actionable driver (little *x*). But these sorts of moves are subject to many business factors and to long cycle times, so they are not usually considered among the critical few direct actions for the purpose of controlling carbon footprint. In scenario planning, however, anticipating the impact of geographic moves on carbon footprint is worthy of discussion. We have seen migrations of energy-intensive datacenter operations from earthquake zones in California to dirty-power states such as Arizona (or to cleaner-power states such as Oregon). If electricity transmission were limited and local feedstock for producing energy became scarce and highly variable from one location to the next, geography would matter even more as a factor for managing carbon footprint.

In the case of water, although it does get transported from place to place, location already matters a great deal. The availability of quality water varies

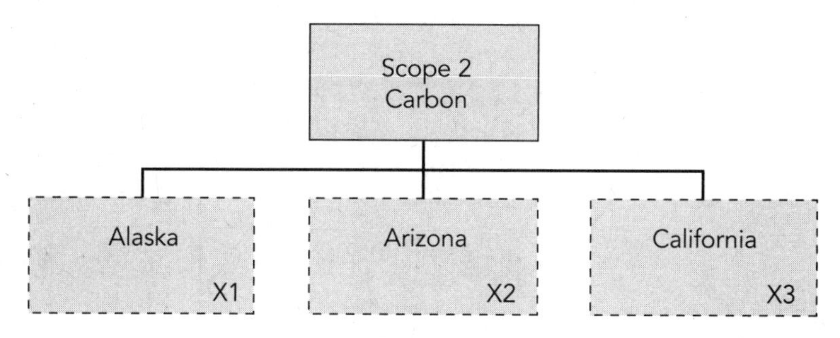

Figure 3-8 Transfer function showing big *X*'s for scope 2 carbon geographies.

from watershed to watershed and from region to region with climate. Local water availability is an important factor for companies in water-intense industries such as energy, food and beverage, pharmaceuticals, and semiconductors. Different regions of the world face water stress for different reasons. According to a recent report by the 2030 Water Resources Group/McKinsey & Company,[7] India's water stress is primarily from growing agricultural demand, China's is from growing industrial demand, and Brazil's is from a more even mix of growing municipal, industry, and agricultural demand. The inevitable tradeoffs between agricultural production and industrial production are of concern to world leaders.

Location decisions for business processes or even for offices housing large groups of employees will more and more need to factor in risks related to regional climate-change impacts, especially in the area of water.

The Transfer Function for Office Water

According to the U.S. Geological Survey, less than 1 percent of all the water on Earth is freshwater available for use by humans (Figure 3-9). Of that water, buildings use 13.6 percent, or 15 trillion gallons per year.

Conceptually similar to carbon-footprint disclosure and management, water footprinting is gaining traction as a way to measure and identify business risks. In 2009, the Water Footprint Network, a nongovernmental organization based in the Netherlands, published the *Water Footprint Manual.* According to this methodology, the total water footprint of a business is defined as

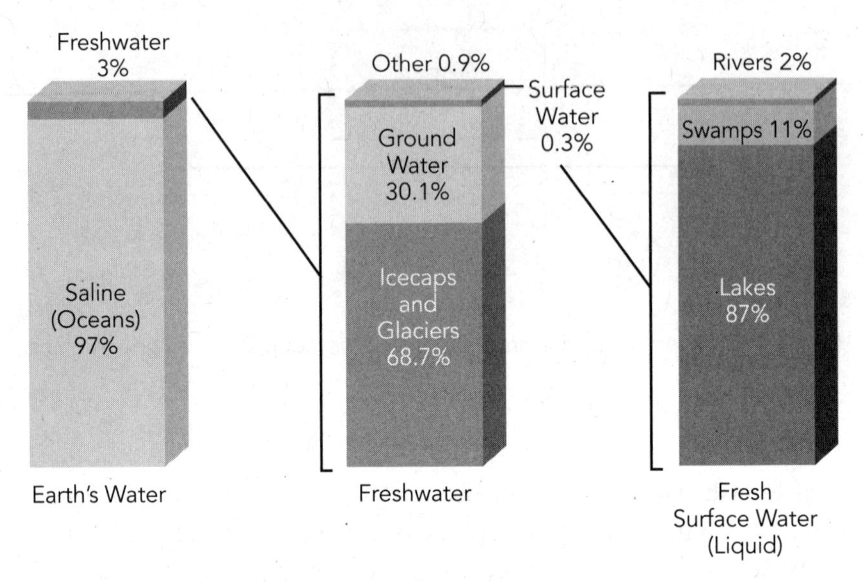

Figure 3-9 Distribution of Earth's water. (*U.S. Geological Survey*, http://ga.water.usgs.gov/edu/earthwherewater.html)

> . . . the total volume of freshwater that is used directly or indirectly to run and support the business. It consists of two main components. The operational (or direct) water footprint of a business is the volume of freshwater consumed or polluted due to its own operations. The supply-chain (or indirect) water footprint of a business is the volume of freshwater consumed or polluted to produce all the goods and services that form the inputs of production of the business.[8]

The *Water Footprint Manual* further explains that for all the business's products and operations, there are three different components (Figure 3-10):

1. *The bluewater footprint.* Bluewater is the amount of fresh surface or groundwater that, under the stewardship of the business, evaporates; is incorporated into products; or is returned to a different watershed or to the same watershed at a different period in time.
2. *The greenwater footprint.* Greenwater is most relevant for agriculture and forest products and is the volume of rainwater consumed during the production process.

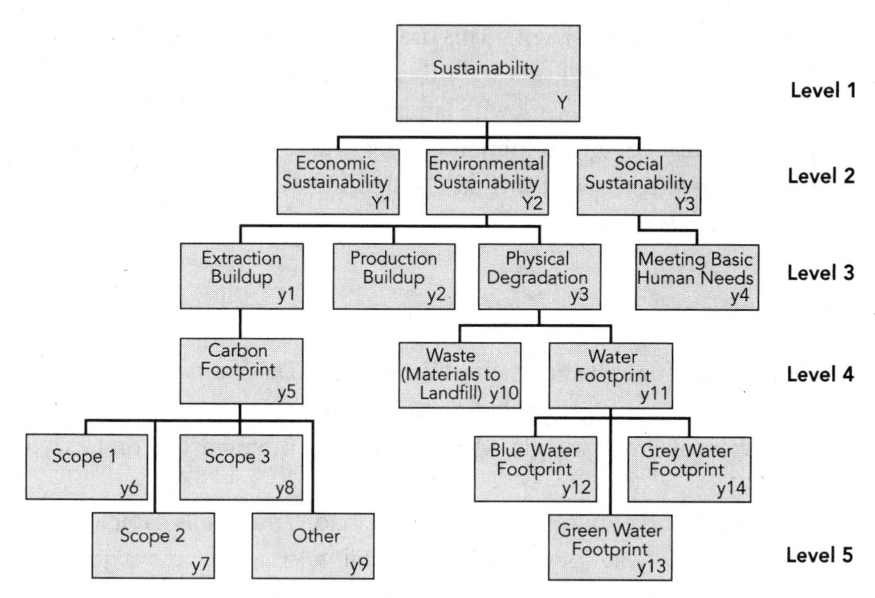

Figure 3-10 The transfer function including water footprint (y11) and its subcomponents (y12–y14).

3. *The graywater footprint.* Graywater relates to pollution and is the volume of freshwater that is required to dilute pollutants from operations to an agreed-on water-quality standard.

For common application across businesses, let's look at the composition of bluewater footprint in commercial office buildings. The water that is used in a typical office building is roughly allocated as follows:

- *Cooling towers (30 to 40 percent of water draw).* This use is driven by equipment, appliances, people, and other heat-producing assets that create a cooling need for the building. To control these loads, use energy-efficient equipment, recommission building systems for optimal performance, install windows with high reflective/insulating value, and follow other best practices in facilities management.
- *Amenities (30 to 40 percent).* This use is driven by human needs in the building, for water for washing, use in restrooms, and drinking. To control these uses, install efficient urinals and toilets, faucet aerators, low-flow shower heads, and water-efficient dishwashers where present in the building. Replace bottled drinking water with tap water filtration (and reusable cups).

- *Irrigation (10 to 30 percent).* This use is driven by the evaporation and transpiration rates of grass and other plants on the building grounds. To control these uses, work with landscape service providers, use native plants instead of grass for landscaping, and install rain-sensor or weather-data-controlled sprinkler systems.
- *Leakage (10 to 30 percent).*

Conclusion

In William McDonough and Michael Braungart's *The Hannover Principles*[9] for sustainable design, principle 6 is "eliminate the concept of waste." In their overview of this principle, McDonough and Braungart go on to say:

> In today's world of trying to be "less bad," materials typically follow a one-way path to the landfill and waste managers intervene here and there to slow down the trip from cradle to grave. . . . Rather than seeing materials as a waste management problem, cradle-to-cradle thinking sees materials as nutrients that cycle through either the biological metabolism or the technical metabolism. . . . [W]aste equals food.

As shown in our examples of the sustainability transfer function, much of the impact of business on the environment is from waste—carbon and wasted energy, wasted materials that end up in a landfill, and wasted water. Business has an opportunity, as McDonough and Braungart inspire in *The Hannover Principles* and in *Cradle to Cradle*, to do more good. With the right approach to product design and to operational management, a company with sustainable development in mind should be able to fashion its products as well as its by-products to be healthy nutrients for the environment. Whatever the business puts in to the biologic system, make sure that it has biologic properties—and likewise for technical systems.

By understanding your own business and environmental sustainability principles, you can build your own company's sustainability transfer function as a means for organizing your program against the critical few drivers of success. The transfer function is an important tool to be able to plan, run, and improve your business. The definition process can be collaborative and engaging for executives and employees in your company.

Individuals should be able to relate their jobs to the transfer function. This sort of collaborative management commands and creates leadership, an important social capital in today's business.

Chapter Summary—Key Points

- The goal of the sustainability transfer function is to define the critical few activities that, when managed, will create results consistent with customer, shareholder, and stakeholder expectations. Mathematically, the transfer function can be thought of as $Y = f(x)$.
- Identifying the transfer function for your company's sustainability efforts should improve the efficiency and effectiveness of planning, running, and improving your program.
- We start the transfer function using the Brundtland definition of sustainability and the commonly accepted three drivers of sustainability: economic, environmental, and social. Using Six Sigma concepts in this book, our goal is to show how economic and environmental concerns can be solved simultaneously.
- To define the next levels of the transfer function, we integrate concepts from the Natural Step system conditions that state
  - In a sustainable society, nature is not subject to systematically increasing
    - Concentrations of substances extracted from the Earth's crust
    - Concentrations of substances produced by society (concentration could be physical or gaseous materials)
    - Degradation by physical means
  - And, in that society,
    - People are not subject to conditions that systematically undermine their capacity to meet their needs
- To further develop the corporate environmental sustainability transfer function, we look at drivers of greenhouse gas emissions, material waste, and water footprint.

Notes

1. Brundtland, *Our Common Future.* Report of the World Commission on Environment and Development. 1987.

2. U.S. Energy Information Agency carbon emission data: www.eia.doe.gov/oiaf/1605/ggrpt/carbon.html.
3. Cisco carbon data: www.cisco.com/web/about/ac227/csr2009/pdfs/CSR_09_Environment_GHG_Emissions.pdf.
4. EPA waste: www.epa.gov/epawaste/nonhaz/index.htm.
5. University of Wisconsin Green Bay e-mail font switch: www.channel3000.com/technology/22954103/detail.html?fbc_channel=1.
6. Xerox solid ink: www.office.xerox.com/solid-ink/solid-ink-benefits/enus.html.
7. International Finance Corporation and McKinsey & Company, *Charting Our Water Future.* 2030 Water Resources Group, New York, NY; November 23, 2009. Available at www.2030waterresourcesgroup.com/water_full/Charting_Our_Water_Future_Final.pdf.
8. Water Footprint Network: www.waterfootprint.org.
9. William McDonough and Michael Braungart, *The Hannover Principles.* New York, NY; William McDonough Architects(1992).

CHAPTER 4

Sustainability Measurement and Reporting

In Chapter 3 we introduced the concept of "big *Y*'s" and overarching sustainability goals for economic, environmental, and social sustainability. With a focus on environmental sustainability, we then discussed the need to develop intermediate- or lower-level goals that will support the attainment of the overarching goal, and we referenced the Natural Step framework, which considers three environmental rules (and one social) to enable a business to be environmentally sustainable:

1. "In order for a society to be sustainable, nature's functions and diversity are not systematically subject to increasing concentrations of substances extracted from the earth's crust." The primary focus here is on the increasing concentrations of gases in our atmosphere from the burning of fossil fuels.
2. "In order for a society to be sustainable, nature's functions and diversity are not systematically subject to increasing concentrations of substances produced by society." A key focus here is on the generation of systematic increases in persistent substances such as DDT, PCBs, and Freon.
3. "In order for a society to be sustainable, nature's functions and diversity are not systematically impoverished by over-harvesting or other forms of ecosystem manipulation." The focus here is on not taking more from the biosphere than can be replenished by natural systems, such as overharvesting of rain forests.

We also identified the fact that one popular measure of success for item 1 above (avoiding increasing concentrations) is the carbon footprint, or the reduction of carbon emissions, and that there are international management and reporting frameworks such as the *Greenhouse Gas Protocol* developed by the World Resources Institute that define a measurement and reporting protocol for greenhouse gas (GHG) emissions. Chapter 3 then went on to describe the transfer-function process that can be used to develop the series of actions (*x*'s) and results (*y*'s) specific to your company that you can use to drive toward overall program success and the achievement of your big *Y*'s.

The focus of this chapter is on describing a variety of international environmental sustainability measurement standards and reporting protocols that have emerged, along with related regulatory and voluntary reporting programs. These reporting and measurement standards, both voluntary and regulated, will need to be incorporated into the big *Y*'s of any corporate environmental sustainability program.

Reporting Overview

Mandatory and voluntary requirements for disclosure of carbon footprint or climate-change-related information have increased steadily over the past decades. The challenge for reporters and their audiences is that climate-change-related measurement and reporting are not entirely standardized, and the data can be difficult to measure and collect. Reporting information for companies is often scattered across their annual reports and separate sustainability reports. Today, there are several emerging trends for measurement and reporting that are described later in this chapter. At the same time, there is an emerging trend toward an *integrated reporting* approach that may provide a solution.

A recent significant development is the convergence of the accounting profession, the sustainability and carbon reporting experts, standard setters, and shareholders to support the creation of a standard platform. The Climate Disclosure Standards Board (CDSB) is a consortium of business and environmental organizations represented by the Carbon Disclosure Project, Ceres, the Climate Group, the Climate Registry, and several more global business associations. It advocates a generally accepted international framework for companies to disclose information about climate-change-related risks and opportunities, carbon footprints, reduction strategies, and their implications for shareholders. The group recommends that disclosed information should focus regulatory and physical climate-change-related risks and GHG emissions. This group created the *Climate Change Reporting*

Framework, which is intended to define what to report as well as how. The final version of the *Climate Change Reporting Framework* was published in September 2010 and is targeted at companies, the accounting profession at large, and regulatory agencies, with a focus on shareholders and investors as the main user group.

The framework recommends two categories for disclosure:

- Overall strategic analysis (short and long term)
- Reporting on risk and governance of climate change and detailed GHG emissions using recognized GHG emissions reporting schemes that include the international *GHG Protocol* reporting standard, the ISO 14064-1 specification, and various national and industry-specific guidelines

Also in August 2010, the Global Reporting Initiative and the Prince of Wales' Accounting for Sustainability Project announced formation of the International Integrated Reporting Committee (IIRC). The IIRC aims to form a consensus among governments, listing authorities, businesses, investors, accounting bodies, and standard setters to establish a globally accepted framework for accounting for sustainability that brings together financial, environmental, social, and governance information in an integrated format.

Benefits and Drivers for Reporting

Performance measurement plays an essential role in strategy development and in evaluating to what extent organizational objectives have been met. Credible GHG accounting will be a prerequisite for participation in GHG trading markets and for demonstrating compliance with government regulations. As evidenced by the significant growth of the Carbon Disclosure Project and the Global Reporting Initiative (GRI), voluntary corporate reporting on climate change has increased in recent years as investor and public interest in climate change has risen dramatically. Key stakeholders in the financial and consumer markets are asking for GHG emissions data and are interested in how companies are contributing to solutions to climate change. Companies are also beginning to ask these questions of their own suppliers (Figures 4-1 and 4-2).

An organization that reports on its sustainability practices is often expected to indicate both where it has succeeded and where it has fallen

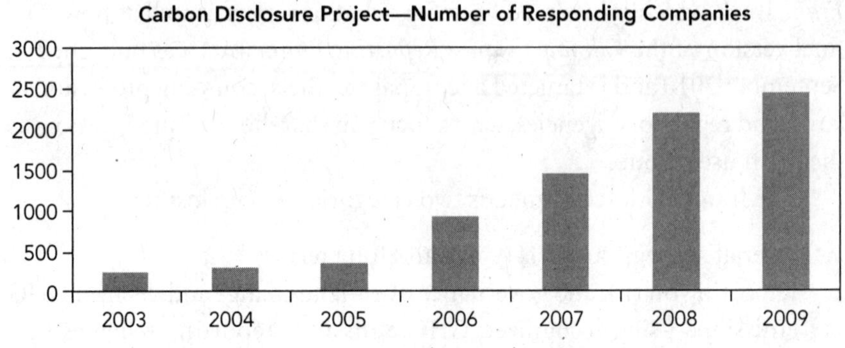

Figure 4-1 Increase in climate-change reporting.

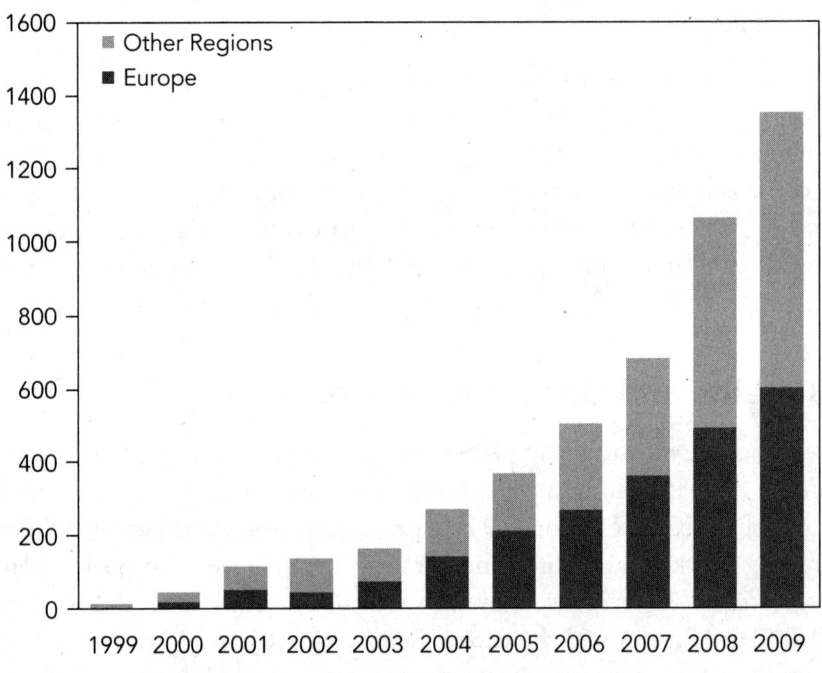

Figure 4-2 Increases in number of GRI reports.

short. This creates an element of reputational risk in the short term but also brings significant long-term benefits, such as

- Better measurement of the organization's triple bottom line
- Stakeholder trust

- Improved risk management
- Improved operational performance

Failure to report on sustainability can increase risk. Companies that do not release sustainability information may appear less transparent than competitors.

Reporting provides the basis to measure emissions, set control targets, motivate employees, and gauge progress toward sustainability objectives. Reporting also creates a public forum in which companies can share best practices and benchmark progress against their peers. There are also benefits that result simply from the internal processes and controls that the companies put in place to collect and analyze the data for reporting, which then can lead to identification of potential operational improvements.

Companies are also engaging their suppliers to improve their procurement policies, and they are engaging their investors and other stakeholders to receive useful feedback on the direction of their strategies and where more progress is desired. Over 3,000 companies in over 60 countries around the world now measure and disclose their GHG emissions and climate-change strategies through the Carbon Disclosure Project (CDP) so that they can set reduction targets and make performance improvements. These data are made available for use to a wide audience including institutional investors, corporations, policymakers and their advisors, public-sector organizations, government bodies, academics, and the public, making the CDP Web site the largest global repository of corporate climate change and GHG emissions data in the world.

Increasingly, companies are finding that transparency and accountability are essential to retaining the trust and confidence of key stakeholder constituencies. Extended to climate change, some companies are finding added benefits through inclusion in investor initiatives, such as the Dow Jones Sustainability Index, and government recognition programs, such as the U.S. Environmental Protection Agency's (EPA's) Climate Leaders program (this program has now been halted) and the EPA ENERGY STAR program. At the same time, a robust disclosure policy can help companies to attract talent from a new, climate-concerned generation. Ultimately, such collaboration around voluntary disclosure programs can lead to necessary innovation and healthy competition. Key drivers of sustainability reporting are described in greater detail below.

Brand

Done properly, reporting on sustainability helps companies to establish a reputation for transparency and to build stakeholder trust. Research conducted by the GRI shows that 82 percent of U.S. companies and 66 percent of European companies cite transparency as the main factor influencing their corporate reputations—a higher percentage than those citing trust, quality of product or services, leadership, or even financial returns.

Messaging around energy efficiency and GHG reductions is becoming increasingly central to the stories that companies want to tell their stakeholders. As a result, such discussions are expanding beyond corporate responsibility reports and CDP submissions into traditional annual shareholder reports, media advertising, and even product labeling. Tesco, the world's third largest food retailer, based in the United Kingdom, began a pilot project to add emissions data to product packaging in April 2008. Other consumer products companies are looking upstream at their supply chain. In September 2007, Walmart announced plans to engage suppliers across seven product categories on climate-change-related reporting. Out of this initiative has grown the CDP's "Supply Chain Customer Demands." Consumers are also raising their voice for corporate action on climate change, making companies increasingly mindful of the branding and reputation risks they face.

Companies are investing in sustainability reporting not only because they think it's good for the planet. Many are doing it because it supports their brand and competitive position.

Track Progress toward Goals

Many companies recognize that just the process of collection and analysis of data for reporting is beneficial in creating a baseline for performance, tracking progress toward goals, and identifying new best practices. Measurement is often the first step toward improvement. A report also serves as an important platform from which a company can communicate with multiple audiences, including potential employees, staff, and shareholders.

Regulatory Compliance

There has been a tremendous increase in legislative and regulatory activity around the globe intended to mitigate climate-change risk. Much of this activity is focused on higher standards for large emitters, such as manufacturers or utility companies, to reduce direct emissions. Other standards are focused on the transportation industry to improve fuel economy, reduce emissions, or accelerate the transition to alternative fuels. Other new laws or standards are focused on the energy efficiency of homes, buildings, and appliances. From a stakeholder perspective, it is important for companies to keep track of these changing standards and regulations in order to maintain and report full compliance.

This has been ever more challenging for U.S.-based companies given the lack of focus on new efficiency or clean energy standards at the federal level. Absent activity at the federal level, many states and municipalities have been passing their own new standards. In the United States, rules are emerging from regional and state initiatives such as the Regional Greenhouse Gas Initiative, the Western Climate Initiative, the Midwestern Greenhouse Gas Reduction Accord, and California's AB32 (the Global Warming Solutions Act). As a result, national and international firms are required to track, comply with, and report against a myriad of new and changing regulations. Legislators, regulators, and governments at the international level are also continuing to introduce laws, regulations, and international accords to mitigate climate-change risks.

Two significant international accords related to this topic are the Kyoto Protocol, which was adopted in Kyoto, Japan, on December 11, 1997, and became effective on February 16, 2005, and the European Union Emissions Trading System (EU ETS), which was launched as an international cap and trade system of allowances for emitting carbon dioxide and other GHGs built on the mechanisms set up under the Kyoto Protocol.

The European Union has long been a driving force in international negotiations that led to agreement on the two United Nations climate treaties, the UN Framework Convention on Climate Change (UNFCCC) in 1992 and the Kyoto Protocol in 1997.

The Kyoto Protocol requires the 15 countries that were EU members at the time (EU-15) to reduce their collective emissions in the 2008–2012 period to 8 percent below 1990 levels. Emissions monitoring and projections show that the EU-15 is well on track to meet this target. In 2007, EU leaders endorsed an integrated approach to climate and energy policy and committed to transforming Europe into a highly energy-efficient, low-carbon economy. They made a unilateral commitment that

Europe would cut its emissions by at least 20 percent of 1990 levels by 2020. This commitment is being implemented through a package of binding legislation.

The EU also has offered to increase its emissions reduction to 30 percent by 2020 on condition that other major emitting countries in the developed and developing worlds commit to do their fair share under a future global climate agreement. This agreement should take effect at the start of 2013, when the Kyoto Protocol's first commitment period will have expired.

Given that this chapter is about reporting and measurement, it should be acknowledged that mandatory reporting itself can be an enabler to allow market forces to prevail and to minimize the need for additional regulation. A parallel example in the appliance industry is that when the U.S. Department of Energy (DOE) required efficiency labels on appliances, an informed public could discern between high-efficiency appliances and those which would have higher operating costs. In a matter of a few short years, the efficiency of the average refrigerator improved by over 25 percent.

The EPA also has been taking steps to regulate GHG emissions. On January 1, 2010, the EPA began, for the first time, to require large emitters of GHGs to collect and report data with respect to their GHG emissions. This reporting requirement is expected to cover 85 percent of the nation's GHG emissions generated by roughly 10,000 facilities. In December 2009, the EPA issued an "Endangerment and Cause or Contribute Finding" for GHGs under the Clean Air Act, which will allow the EPA to craft rules that directly regulate GHG emissions.

Shareholder Pressure/Securities and Exchange Commission (SEC) Guidance

Investors in the United States have been calling for more standardized climate disclosure in securities filings. In February 2010, the SEC published interpretative guidance regarding its disclosure requirements related to climate-change risk. Issued in response to petitions from institutional investors, the guidance does not amend any existing disclosure requirements or create any new ones, but it does signal that companies should maintain a heightened awareness of climate-change risk when preparing disclosures for SEC filings. Reporting regularly on sustainability could be one means of maintaining such awareness.

Regular reporting also could prevent companies from violating stricter truth-in-advertising standards. In October 2010, the U.S. Federal Trade Commission (FTC) issued new guidance that puts pressure on companies to substantiate claims made in a sustainability report, such as statements that products are "recyclable" or "carbon neutral."

The SEC provided guidance in four categories:

1. *Impact of legislation and regulation.* Companies will have to consider whether the impact of existing climate laws and regulations are material and in "certain circumstances" whether pending laws and regulations are also material. The materiality of pending regulations and laws could be a matter of contention given the uncertainties about the future of both federal agency rule makings and legislation, as well as state initiatives.
2. *Impact of international accords.* Companies will be required to consider (and disclose when material) business impacts of "international accords and treaties relating to climate change." Here as well, the uncertainties surrounding the future of the Kyoto Protocol and the Copenhagen Accord could complicate disclosures.
3. *Indirect consequences of regulation or business trends.* Legal, technological, political, and scientific developments regarding climate change may create new opportunities or risks for companies. Demand for carbon-intensive goods may go down, and the cost of certain equipment, fuels, and material may increase. The SEC thus will require disclosure of "actual or potential indirect consequences [a company] may face" from climate-change laws or policies.
4. *Physical impacts of climate change.* Companies will be required to evaluate the "actual and potential material impacts of environmental matters" on their business. Again, the ultimate guidelines may provide more detail about the expectations that the SEC has for disclosures and, in particular, how it believes public companies should address areas for which there are significant uncertainties.

Reporting and Measurement Standards/Protocols

During the past decade, the ideas of transparency and accountability in environmental and sustainability performance have gained traction with most large global corporations. Voluntary sustainability reporting has emerged as part of this trend. The GRI rapidly has become the leader among voluntary worldwide performance reporting programs, whereas other reporting programs and measurement standards also have emerged. Several challenges remain in finding a common language for GHG accounting, especially as voluntary and mandatory disclosure requirements intersect—often with important legal implications. Yet, despite these challenges, companies are seeing the benefits of this disclosure. In this section we will describe some of the leading standards for measurement and reporting.

The Greenhouse Gas Protocol

The global leading standard for the measurement or accounting (versus reporting) of GHGs is the *Greenhouse Gas Protocol.* Unlike for financial accounting and reporting, there are no "generally accepted accounting and reporting practices" for corporate GHG emissions. The *Greenhouse Gas Protocol Corporate Standard* (GHGP) provides standards and guidance for companies and other organizations preparing a GHG emissions inventory. The reporting guidelines include the six GHGs covered by the Kyoto Protocol—carbon dioxide (CO_2), methane (CH_4), nitrous oxide (N_2O), hydrofluorocarbons (HFCs), perfluorocarbons (PFCs), and sulfur hexafluoride (SF_6).

The Greenhouse Gas Protocol (GHG Protocol) is the most widely used international accounting tool for business leaders to understand, quantify, and manage GHG emissions. The *GHG Protocol* provides the accounting framework for nearly every GHG standard and program in the world, as well as hundreds of GHG inventories prepared by individual companies. The *GHG Protocol*'s toolset enables companies to develop comprehensive and reliable inventories of their GHG emissions, and it is the accepted international standard, having been widely implemented by companies and industry associations. It also has been adopted by the International Organization for Standardization (ISO) as the basis for its ISO 14064, as well as the GRI and CDP. The *GHG Protocol* is a significant step toward the creation of generally accepted GHG accounting and reporting practices.

While the *GHG Protocol* helps companies to frame where emissions may be coming from, it also assists in determining the correct organizational and operational boundaries for measurement. The operational boundaries are divided into scope 1, scope 2, and scope 3, as described below. For organizational boundaries, the financial reporting standards are very clear, as in defining which subsidiaries you should include as part of an organization. However, with GHGs, the challenge is that reporting is not necessarily restricted to emissions for which the company has legal liability or direct control. This process of defining the environmental limits of an organization is referred to as *boundary setting* and is still a new and emerging area.

The operational boundaries and scope categories begin first by defining direct versus indirect GHG emissions:

- *Direct* GHG emissions are emissions from sources that are owned or controlled by the reporting company, for example, emissions from factory stacks, manufacturing processes and vents, and company-owned/controlled vehicles.
- *Indirect* GHG emissions are emissions that are a consequence of the activities of the reporting company but occur from sources owned or controlled by another company, for example, emissions from the production of purchased electricity, contract manufacturing, employee travel on scheduled flights, and occurring during the product use phase.

The concept of *scope* was introduced to help delineate direct and indirect emissions sources, improve transparency, and provide utility for different types of organizations with different needs and purposes. Three scopes are defined for GHG accounting and reporting purposes. The *GHG Protocol* recommends that companies account for and report on scopes 1 and 2 at a minimum.

Scope 1 accounts for direct GHG emissions from sources that are owned or controlled by the reporting company. Scope 1 emissions are principally the result of the following activities:

- Production of electricity, heat, or steam
- Physical or chemical processing
- Transportation of materials, products, waste, and employees, for example, use of mobile combustion sources such as trucks, trains, ships, airplanes, buses, and cars
- Fugitive emissions (intentional or unintentional releases), for example, equipment leaks from joints or seals, methane emissions from coal mines, and HFC emissions during the use of air-conditioning equipment

Scope 2 accounts for indirect emissions associated with the generation of imported/purchased electricity, heat, or steam. For many companies, electricity usage represents one of the most significant opportunities to reduce GHG emissions. Companies can reduce their use of electricity and/or use it more efficiently by investing in energy-efficient technologies. Additionally, emerging green power markets enable some companies to switch to less GHG-intensive electricity suppliers.

Companies also can install an efficient cogeneration plant on site to replace the import of more GHG-intensive electricity from the grid. Scope 2 facilitates the transparent accounting of such choices.

Scope 3 allows for the treatment of other indirect emissions that are a consequence of the activities of the reporting company but occur from sources owned or controlled by another company, such as

- Employee business travel
- Transportation of products, materials, and waste
- Outsourced activities, contract manufacturing, and franchises
- Emissions from waste generated by the reporting company when the point of GHG emissions occurs at sources or sites that are owned or controlled by another company, for example, methane emissions from landfill waste
- Emissions from the use and end-of-life phases of products and services produced by the reporting company
- Employees commuting to and from work
- Production of imported materials (Figures 4-3 and 4-4)

The Global Reporting Initiative (GRI)

The GRI is a multistakeholder-governed, independent, nongovernmental organization collaborating with companies and other organizations throughout the world to provide a generally accepted framework for sustainability reporting. It has developed the world's most widely used sustainability reporting framework. This framework sets out the principles and processes that organizations can use to measure and report their economic, environmental, and social performance. More than 2,000 companies and organizations, including many of the world's leading brands, have declared their voluntary adoption of the GRI guidelines. The GRI aims

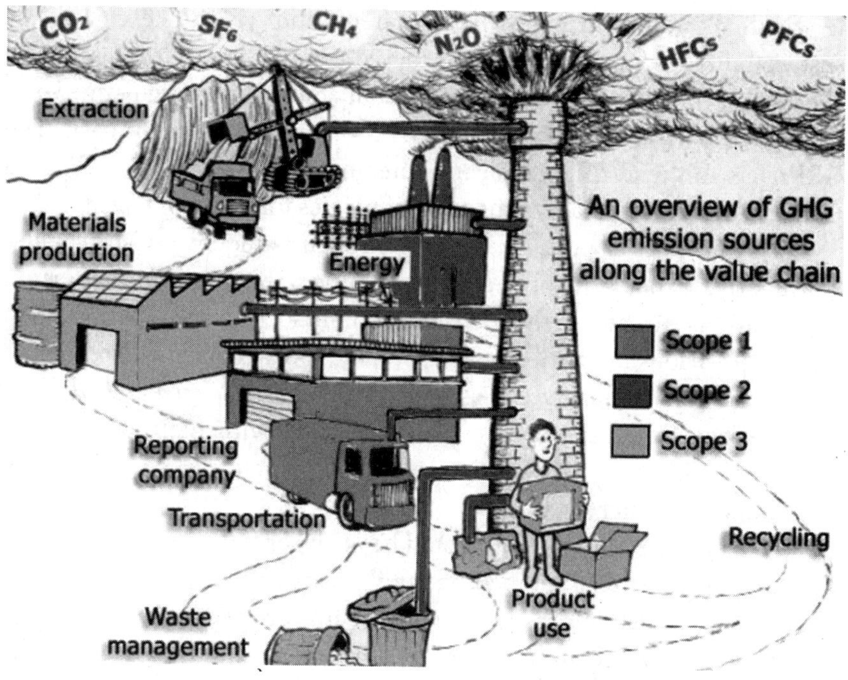

Figure 4-3 Overview of GHG emission sources.

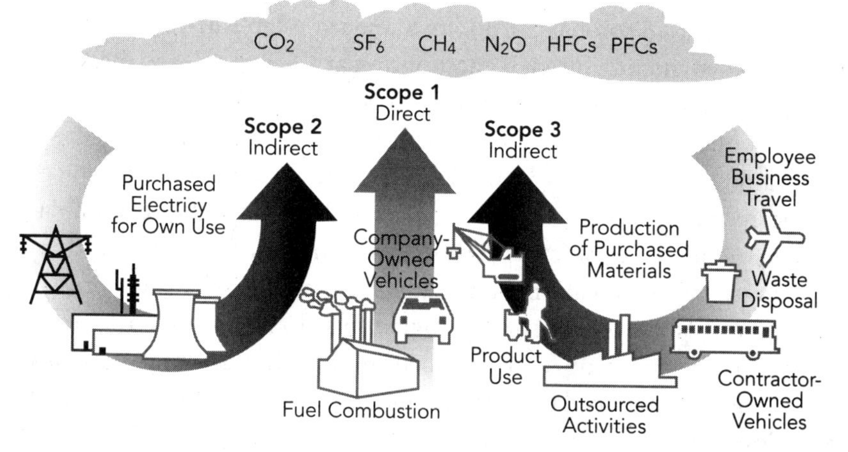

Figure 4-4 Scope 1, 2, and 3 emissions.
(Source: New Zealand Business Council for Sustainable Development.)

to create conditions for the transparent and reliable exchange of sustainability information. This is achieved through the development and continuous improvement of the GRI Sustainability Reporting Framework. This framework allows companies to be compared on a similar basis. The GRI refers to the *GHG Protocol* and Intergovernmental Panel on Climate Change (IPCC) for data sets and methodologies for its reporting.

The guidelines have been developed through a unique multistakeholder consultative process involving representatives from various reporting organizations. To cater to the specific corporate social responsibility (CSR) aspects of various industries, sector-specific report supplements also have been released. The most recent addition is the *Construction and Real Estate Sector Supplement* (CRESS). Several of its environmental performance indicators ask for climate-change-related information such as building energy and CO_2 intensity.

On first inspection, the GRI could be mistaken as an alternative to the *GHG Protocol* or even a competitor, but it is actually more concerned with specific metrics and measurement-reporting processes and not GHG measurement and accounting. It is not competitive with the *GHG Protocol* per se but rather complementary. Companies can implement the *GHG Protocol* within the GRI framework (Figure 4-5).

The Climate Registry

While the GHG Protocol Corporate Standard established internationally recognized standards for GHG accounting at the entity level, the Climate Registry is designed to ensure that GHG emissions data are made available to the public through annual reports posted on the registry's Web site. This program covers 39 states in the United States, 7 Canadian provinces, 6 Mexican states, 3 Native American tribes, and the District of Columbia.

The Climate Registry sets consistent and transparent standards for the measurement, verification, and public reporting of GHG emissions throughout North America in a single unified registry. The registry is a nonprofit organization that supports both voluntary and mandatory reporting programs and provides meaningful information to reduce GHG emissions.

The registry is committed to:

- Using best practices in GHG emissions reporting

Deciding What to Disclose: Guidelines from GRI

Many companies follow the GRI framework when deciding what to include in their sustainability reports. The framework starts with a series of principles that organizations can use to judge whether a particular piece of information merits inclusion in their sustainability reports. The principles are materiality, stakeholder inclusiveness, sustainability context, and completeness.

Materiality—Information in the report should reflect the company's most significant impacts to society and the environment. Material issues can be those that affect the organization's financial position in the short term, but can also extend to factors with longer-term implications. Determining what is material requires that the organization assess its "overall mission and competitive strategy, concerns expressed by stakeholders, broader social expectations, and the organization's influence on upstream (e.g., supply chain) and downstream (e.g., customers) entities."

Stakeholder inclusiveness—Reports should respond to stakeholders' "reasonable expectations and interests." In this context, stakeholders are any individuals or communities likely to be significantly affected by what the organization does, or whose actions are likely to affect the organization's ability to carry out its business strategy and achieve its goals.

Sustainability context—The purpose of a sustainability report is to show how an organization is helping to improve (or at least halt the deterioration of) environmental, social, or other conditions over the long term. Reporting on isolated or narrowly local instances of improvement fails to meet this objective. For example, an organization reporting on the benefits it provides to employees could put those benefits in context by presenting them "in relation to nationwide minimum and median income levels and the capacity of social safety nets to absorb those in poverty or those living close to the poverty line."

Completeness—Reports should reflect significant impacts of the business and enable stakeholders to assess its performance in the reporting period.

Source: GRI Sustainability Reporting Guidelines, Version 3.0

Figure 4-5 GRI guidelines.

- Establishing a common data infrastructure for voluntary and mandatory reporting and emissions-reduction programs
- Minimizing the burden on reporters and members
- Providing an opportunity for reporters to establish an emissions baseline and document early action
- Developing a recognized platform for credible and consistent GHG emissions reporting

The Carbon Disclosure Project (CDP)

The CDP is an independent not-for-profit organization holding the largest database of primary corporate climate-change information in the world.

Thousands of organizations from across the world's major economies measure and disclose their GHG emissions and climate-change strategies through CDP. The CDP puts this information at the heart of financial and policy decision making. The CDP collects and distributes climate-change information, both quantitative (emissions amounts) and qualitative (risks and opportunities), on behalf of 534 institutional investors. Over 3,000 companies globally now report to the CDP, including over 500 U.S. companies. Sixty-eight percent of the companies that responded to the CDP's investor requests for information made their reports available to the public. After 10 years of existence and its eighth year of reporting, 534 institutional investors representing $64 trillion of assets under management now back the CDP. Eighty-two percent of the world's 500 largest listed companies have provided insight into carbon emissions, regulatory, and physical risks and opportunities. In contrast to the GRI, the CDP clearly focuses on the carbon and climate-change-related risks and opportunities of a company.

The CDP leverages its data and processes by making its information requests and responses from corporations publicly available, helping to support the activities of policymakers, consultants, accountants, and marketers.

The GRI and the CDP now have released a linkage document that demonstrates the growing collaboration between the two organizations. The 16-page document outlines how CSR reporters can use or adapt the same data for both reporting processes. As more companies become transparent in disclosing their environmental, social, and governance (ESG) impacts to their shareholders, greater standardization is needed. Both organizations will use the document as a framework for improving their respective guidelines and questionnaires. The CDP amends its annual questionnaire annually and can incorporate GRI standards into its framework. Likewise, GRI is working on a new iteration of its framework and is exploring a new set of guidelines as it pushes for greater ESG and integrated reporting.

Both organizations are learning from each other. For example, the CDP requires more detailed information on GHG and carbon emissions than the GRI. The GRI is more rigorous in parsing out a company's energy consumption throughout its supply chain, and the CPD does not request any data about a firm's indirect energy consumption.

Finally, the CDP asks its participating firms pointed questions about how climate change could affect their operations, quite similar to the SEC's recent advisory to public companies regarding the materiality of the risks involved with global warming. The GRI, which is advocating integrated reporting, does not suggest such disclosure in its guidelines.

Even with all this activity related to the development of measurement and reporting standards, there is a need for greater standardization. While many companies have adopted the GHG Protocol, there still exists a wide range of methodologies for calculating and reporting GHG emissions. This is especially true of companies tracking scope 3—or indirect—emissions and calculating emission offsets. Such lack of comparability has made it difficult to benchmark company performance in investment analyses. To help work toward better standardization, the CDP has been appointed as the secretariat for the Carbon Disclosure Standards Board, which was convened at the World Economic Forum in 2007 to jointly advocate a common reporting framework.

International Standards Organization (ISO)

The ISO has developed a set of standards in the 1400 series designed to help organizations manage their overall environmental impact. The measurement and reporting of GHG emissions are specifically addressed in ISO 14064, which specifies principles and requirements at an organization level for the quantification and reporting of GHG emissions and removals (including requirements for the design, development, management, reporting, and verification of GHG inventories).

Process and/or Tools

While the reporting standards provide a basic framework for accounting and reporting of GHG emissions at the corporate level, a number of technical accounting and reporting decisions are left to its users to make, based on their individual or programmatic goals. An important starting point for a company contemplating GHG performance measurement is to understand where the measures link with the company's business drivers and what their relevance to company performance will be. This also will encourage buy-in to the system from employees and senior management, who may be faced with a range of competing objectives.

The measurement and reporting standards that we have discussed thus far do not adequately capture the immense corporate commitment in terms of time and resources necessary to gather the data and compile a report that meets these standards. Again, the *GHG Protocol* has issued some guidance that suggests a logical sequence of activities beginning with the establishment of business goals:

1. Consider business goals.
2. Consider GHG accounting principles.
3. Define organizational boundaries.
4. Define operational boundaries.
5. Select base year.
6. Identify emissions sources.
7. Calculate emissions.
8. Verify inventory.
9. Report emissions.
10. Establish GHG reduction target.

Each of these steps is described briefly in the following paragraphs.

Consider Business Goals

1. GHG risk management:
   - Identifying GHG risks and reduction opportunities in the value chain
   - Setting internal targets, measuring and reporting progress
   - Identifying cost-effective reduction opportunities
   - Developing process/product innovations
   - Internal/external benchmarking
2. Public reporting/participation in voluntary initiatives:
   - Stakeholder reporting (e.g., GRI)
   - Voluntary nongovernmental organization (NGO) programs (e.g., Climate Neutral Network, WWF Climate Savers Program, Environmental Resources Trust)
   - Voluntary government programs (e.g., Canadian Climate Change Voluntary Challenge and Registry, Australian Greenhouse Challenge Program, California Climate Action Registry)
   - Ecolabeling and certification

3. GHG markets:
   - Buying or selling emissions credits
   - Cap-and-trade allowance trading programs (e.g., UK Emissions Trading Scheme)
4. Regulatory/government reporting:
   - Directives (e.g., European Integrated Pollution Prevention and Control Directive, European Pollutant Emission Register)
   - Reporting under national or local regulations (e.g., Canadian National Pollutant Release Inventory)
   - Carbon taxes
   - Baseline protection

Consider GHG Accounting Principles

1. *Relevance:* Define boundaries that appropriately reflect the GHG emissions of the business and the decision-making needs of users.
2. *Completeness:* Account for all GHG emissions sources and activities within the chosen organizational and operational boundaries. Any specific exclusion should be stated and justified.
3. *Consistency:* Allow meaningful comparison of emissions performance over time. Any changes to the basis of reporting should be clearly stated to enable continued valid comparison.
4. *Transparency:* Address all relevant issues in a factual and coherent manner based on a clear audit trail. Important assumptions should be disclosed and appropriate references made to the calculation methodologies used.
5. *Accuracy:* Exercise due diligence to ensure that GHG calculations have the precision needed for their intended use, and provide reasonable assurance on the integrity of reported GHG information.

Define Organizational Boundaries

The general guidance with respect to GHG reporting is that corporate organizational boundaries should be consistent with the organizational boundaries that have been used for financial reporting purposes. These factors come into play with joint ventures and partially owned entities or facilities. Again, the *GHG Protocol* provides detailed guidance here.

Define Operational Boundaries

Operational boundaries refers to the categorization of emissions as direct or indirect. Direct emissions are from sources that are owned or controlled by the reporting company, such as from factory stacks or company-owned vehicles. Indirect emissions are a consequence of the activities of the reporting company but occur from sources owned or controlled by another company, such as purchased electricity or employee travel. To help to delineate direct and indirect emission sources, the *GHG Protocol* introduced the concept of scope 1, scope 2, and scope 3 emissions discussed previously.

Select Base Year

The *GHG Protocol* recommends setting a historical performance baseline for comparing emissions over time. If you intend to participate in a voluntary GHG reductions program or a GHG emissions trading scheme, it is important to check to determine whether it has any specific rules governing the establishment of base-year emissions. The UK Emissions Trading Scheme, for example, specifies that the baseline will be the average emissions in the three years up to and including 2000. Companies should choose a base year for which verifiable data are available and should develop a base-year emissions-adjustment policy to clearly articulate the basis for making any adjustments. The *GHG Protocol* provides detailed guidance on establishing rules for base-year adjustments that are due to organizational structural changes, transfer of ownership, and so on.

Identify Emissions Sources/Calculate Emissions

Once the organizational and operational boundaries have been established, companies can begin to collect the raw data and perform the necessary calculations to report on GHG emissions. The key steps are to:

- *Identify GHG sources.* The *GHG Protocol* describes these in four overall categories—stationary combustion, mobile combustion, process emissions, and fugitive emissions—and then further classifies them in the scope 1, scope 2, and scope 3 categories discussed previously.
- *Select an emissions calculation approach.* For some emissions, this is merely a task of calculating the amount of fuel consumed and then

applying standard emission factors for the fuel. It is more complicated for process emissions, which may be based on chemical mass-balance equations. Again, the *GHG Protocol* provides detailed guidance.

- *Collect activity data and choose emissions factors.* This is the step to collect actual emissions data. For consumed electricity, this may be relatively simple by using utility meter data, but if you are factoring in your electricity consumption in leased office space and are not directly metered, you may have to estimate.
- *Apply calculation tools.* The *GHG Protocol* has developed calculation tools that are available on the Web site. Also, many of the large accounting software firms and other new startup firms have developed GHG accounting programs.

Verify Inventory

The *GHG Protocol* also provides guidance for ensuring and maintaining the quality of the collection, calculation, and reporting processes, including considerations for when to use and state assumptions.

Reporting Emissions

One of the final steps is to actually report on the emissions. The *GHG Protocol* provides some guidance in this area, but much more detailed guidance is included in the GRI and CDP standards.

Establish a GHG Reduction Target

With the assumption that all of the preceding was performed in order to establish a baseline measurement and report, it may now be necessary to establish future goals or targets for reductions. Some factors to consider include the notion of absolute reduction targets irrespective of growth of the business or the use of ratio indicators.

There are two principal aspects of GHG performance that are of interest to management and stakeholders. One concerns the overall GHG impact of a company or an organization—that is, the absolute quantity of GHG emissions. The other concerns the performance in reducing GHG emissions, measured in ratio indicators. Ratio indicators provide information on relative performance and are described below:

1. *Productivity/efficiency ratios.* Productivity/efficiency ratios express the value or achievement of a business related to its GHG impact. Increasing efficiency ratios reflect a positive performance improvement. Examples of productivity/efficiency ratios include resource productivity (e.g., sales per GHG) and process ecoefficiency (e.g., production volume per amount of GHGs).
2. *Intensity ratios.* Intensity ratios express the GHG impact per unit of activity or unit of value. A declining intensity ratio reflects a positive performance improvement. Many companies historically tracked environmental performance with intensity ratios. Examples of intensity ratios include ton of CO_2 emissions per unit of output or per person or per occupied square feet of offices.
3. *Percentages.* A percentage indicator is a ratio between two similar issues (with the same physical unit in numerator and denominator). Examples of percentages that can be meaningful in performance reports include current GHG emissions expressed as a percentage of base-year GHG emissions.

Examples of Voluntary Reporting Initiatives

The previous sections have described various global reporting protocols and the necessary steps to gather and report your data in compliance with these standards. Once you have completed the hard work of collecting and reporting your greenhouse gas emissions data, there is the question of what to do with this information. Besides the obvious answer to use the data as part of a management program to improve your performance, there are other corporate benefits for public disclosure in terms of branding and awareness as discussed previously. In addition to the Carbon Disclosure project, there are other voluntary reporting initiatives that should be considered as summarized in Figure 4-6.

Sustainability Investment-Rating Agencies

Related to the topic of performance measurement and reporting is the emergence of sustainability investment-rating agencies. The emergence of

Voluntary Initiative	Focus Entity (Company) or Project	Gases Covered	Operation Boundaries (Direct/Indirect Emissions)	Organizational Boundaries
Australian Greenhouse Challenge	Entity (Australian operations)	Six Kyoto gases	Scopes 1 and 2	Distinguishes between GHGs the entity controls and those it influences, reductions from influenced activities are reported separately
California GHG Registry	Entity (See legislation for details)	Six Kyoto gases	Scopes 1 and 2 required, Scope 3 to be determined	Consistent with GHG Protocol
Canada Climate Change Voluntary Challenge and Registry	Entity	CO_2 required, other Kyoto gases optional	Flexible (Scopes 1, 2, or 3)	100 percent of emissions from operated facilities
Environmental Resources Trust—GHG Registry	Entity and project (with verifiable baseline)	Six Kyoto gases	Scope 1	Case-by-case basis, depending upon ownership structure and operating control
US EPA Climate Leaders Initiative	Entity (US operations required, global operations optional)	Six Kyoto gases	Scopes 1 and 2 required, Scope 3 optional	Consistent with GHG Protocol
US Voluntary Reporting on GHG (1805b Program)	Entity or project (US and international operations of any US company)	Six Kyoto gases and ozone precursors	Flexible (Scopes 1, 2 or 3)	Identification of other potential reporters to same emission reduction required
World Wildlife Fund Climate Savers Program	Entity	Energy related CO_2, other gases on a negotiated basis	Scopes 1 and 2 required, Scope 3 optional	Consistent with GHG Protocol

Figure 4-6 Examples of voluntary reporting initiatives.

several sustainability-driven financial indexes, including the Dow Jones Sustainability Index, Ethibel, SERM, and FTSE4GOOD, has helped investors to make informed decisions by evaluating companies across a broad range of economic and environmental criteria. When evaluating companies for inclusion, these indexes consider policies the company has in place, how it reports on environmental impact, and whether it monitors its suppliers. There are two main global sustainability indices that analyze companies based on their ESG performance: the FTSE4GOOD Index and the Dow Jones Sustainability Index.

The FTSE4GOOD Index series is published by the *Financial Times*/London Stock Exchange (FTSE) group and EIRIS (an ESG-focused not-for-profit investment research company based in London). One of the five groups of ESG performance criteria addresses climate-change mitigation and adaptation, which requires information on the volume of GHG emissions and sector-specific CO_2 intensity metrics, goals, and targets, and so on. For the time being, the FTSE group only analyzes industries with the highest levels of associated emissions and thus the greatest need to address this issue (e.g., mining, aluminium and steel production, electric power generation, building materials, and airlines). The FTSE group publishes not only an overall sustainability index of which climate change is a component but also a climate-change-focused index called the *FTSE Carbon Index Series.*

The Dow Jones Sustainability Index series is published by Sustainable Asset Management (SAM), a Zurich-based investment firm and subsidiary of Rabobank/Robeco. It launched in 1999 in collaboration with Dow Jones Indexes. A total of 2,500 of the world's biggest companies in all industry sectors are analyzed across 58 industry sectors, and a leader is identified for each of the 19 supersectors, such as the GPT Group for Real Estate in 2010. There is only one question regarding the environment that covers GHG emissions and energy consumption.

Buildings

On a global basis, buildings and the energy they consume account for up to 40 percent of total GHG emissions. For service companies that do not have manufacturing operations or are not in the transportation industry, their carbon footprint is comprised almost exclusively of the emissions from the offices they occupy and emissions from travel, both employee commutes and business travel. As a result, significant focus has been placed on energy and carbon measurement, benchmarking, rating/certification, and reporting related to residential and commercial buildings.

Building rating systems fulfill a number of important roles. While they essentially provide a standard for what systems, materials, and strategies can help to make a building green, they also are key tools for using the market to increase demand for high-performance buildings. They provide a means for a building owner or tenant to assess one building relative to another. They also enable organizations that are working to effect market transformation to use building rating systems as a tool for specifying minimum performance levels and create industry standards that are above and beyond what is required by code. These rating systems also enhance overall understanding of how buildings contribute to GHG emissions and help to create a broader understanding of the design and operational changes that can be implemented to improve performance.

Since introduction of the Building Research Establishment Environmental Assessment Method (BREEAM) in the United Kingdom in 1990, there has been a proliferation of these rating systems around the world, and they continue to evolve. Since 2000, the number of environmental assessment methodologies around the world has been increasing rapidly. BREEAM was the first system (launched in 1990) to offer an environmental label for buildings. There are now a number of different green building rating systems around the world, most of which have been based on or inspired by BREEAM. The most prevalent system is the Leadership in Energy and Environmental Design (LEED) Program developed and managed by the U.S. Green Building Council. It is by far the most common rating and certification system in the United States and is now being adopted in many countries around the world. The LEED Program has many different certification types focused on different asset classes, such as residential, retail, or commercial office, and distinctions between new construction, existing buildings, and tenant spaces.

At the same time, the investment community has been seeking ways to render building energy performance more transparent with a focus on the preceding rating systems and on more intense and detailed energy-efficiency measurement, benchmarking, and labeling programs. Two types of energy-efficiency labeling systems have emerged: private labels, such as ENERGY

STAR in the United States, and mandatory labels, such as Energy Performance Certificates for buildings due to be sold or leased in the European Union.

As a consequence, buildings may represent different levels of attractiveness to both investors and occupiers, who will be better able to understand potential energy costs for space they are using. The ENERGY STAR system in the United States is a free system that allows building owners and managers to enter data about the energy consumption of their buildings, hours of operation, number of occupants, location, and other building characteristics. The system then compares each building's energy performance with the performance of "similar" buildings in the local market. Each building receives an ENERGY STAR rating from 1 to 100 based on one year's energy-consumption data. ENERGY STAR normalizes the data for weather variations and basic operating conditions. Annual energy consumption in buildings can vary up to 30 percent depending on local weather. Buildings that achieve an ENERGY STAR score of 75 or greater (indicating that they are in the top quartile in their market) are able to achieve the ENERGY STAR label and certification for the building (Figure 4-7).

Rating System	Provider(s)	Country
BEAM	HK BEAM Society	Hong Kong
BREEAM	Building Research Establishment	United Kingdom
CASBEE	Japan Green Building Council	Japan
DGNB	German Sustainable Building Council	Germany
ECB	Building Living Dialogue Programme	Sweden
Estidama	Abu Dhabi Urban Planning Council	Unite Arab Emerates
Green Globes/ Go Green	GBI/BOMA Canda	United States/Canada
Green Star	Green Building Council of Australia, New Zealand Green Building Council, Green Building Council of South Africa	Australia, New Zealand, South Africa
GRIHA	Tata Energy Research Institute	India
HQE	Association pour la Haute Qualité Environnementale	France
LEED	U.S. Green Buidling Council	United States
NABERS	Australian Government	Austrialia
Three Star	Ministry of Construction/Ministry of Housing and Urban-Rural Development	China

Figure 4-7 Common building rating systems around the world.

Chapter Summary—Key Points

- Incorporate these measurement and reporting standards into the sustainability transfer function for your company.
- Mandatory and voluntary requirements for disclosure of environmental information have increased steadily over the past decades.
- Developments in integrated reporting and in standardizing climate-change reporting are gaining traction.
- An organization that reports on its sustainability practices is often expected to indicate both where it has succeeded and where it has fallen short.
- Reporting provides the basis to measure emissions, set control targets, motivate employees, and gauge progress toward sustainability objectives. Reporting also creates a public forum in which companies can share best practices and benchmark with peers.
- We cover similarities and differences in reporting concepts from the Kyoto Protocol, the EU Emissions Trading System, the UN Framework Convention on Climate Change, the Global Reporting Initiative, the Carbon Disclosure Project, and other protocols.
- Some of the major sustainability investment-rating agencies are the Dow Jones Sustainability Index, Ethibel, SERM, and FTSE4GOOD.
- Common rating systems for buildings include LEED, BREEAM, and ENERGY STAR.

CHAPTER 5

Transformational Change and the Power of Teams

As we review sustainability initiatives that have failed to achieve their desired results, two distinct but closely aligned factors stand out. They are the failure to plan or execute an effective change-management strategy appropriately and the failure to leverage the use of teams as a part of that overall change-management strategy. We find this fact to be particularly puzzling because so much has been written about the importance of change-management strategy and the overall impact that teams can have in accelerating change.

Much can be drawn from our overall Six Sigma body of knowledge to understand how to design and execute a change-management strategy and how to leverage teams in the acceleration of that change. By design, starting with the very first projects at Motorola, Six Sigma implementation has been a team-based sport. Structures and processes are designed to ensure team success. Too often these structures are ignored, and the teams fail. But it is useful to review the structures and processes that have supported the success of Six Sigma teams that do make an impact.

In this chapter we will review our findings of why so many green project teams have failed to make an impact. We then will compare those failure modes with the structures and processes that have enabled success in many Six Sigma program implementations. We will explain how this Six Sigma high-performance-teams model fits into an overall change-acceleration strategy to support sustainability initiatives. Finally, we will apply these principles to driving positive change across an organization and across all the stakeholders in a sustainability initiative.

Why Green Project Teams Fail

Most managers understand and embrace the concept of teams and the importance of well-functioning teams to drive a complex sustainability initiative. Unfortunately, it takes more than T-shirts and team parties to make teams successful. As we reviewed team-based sustainability initiatives, we found that some basic misconceptions about how to organize and manage teams led to the failure of the initiatives. Many managers practice principles related to the concepts of *empowerment* that failed miserably in the 1990s but continue to find their way into popular reading.

The basic teachings of *empowerment theory* are that if we give people and teams a good understanding of the situation, a certain level of authority, resources with which to take action, and an environment where they can think creatively, they will be motivated, excited about their work, and essentially do the right things for the organization. Managers responsible for these teams send them on their way and then light a candle and hope for good things to happen. In reality, these self-directed teams have certain deficiencies built into their design. Each of these deficiencies is discussed below.

Uncertainty of Purpose, Lack of Goal Clarity

Our analysis indicates that green project teams are often launched with very broad directives:

- Do our part to save the earth.
- Reduce energy consumption across the company.
- Engage all employees in our green initiative.

The result is that team members are empowered to interpret each directive according to their own needs and perspectives. While this is well-intentioned empowerment, these broad directives rarely lead to meaningful impact. Teams will invest inordinate amounts of time debating their purpose and setting abstract, unachievable goals. Because of this, teams often die in their formative stage. They will have T-shirts from the kick-off but nothing to show for their work.

Narrow Focus

The flip side of the lack-of-clarity failure mode is the too-narrow-of-a-focus failure mode. In this failure mode, teams develop very specific, very narrow objectives without looking for root causes or opportunities to leverage their efforts. An example is a team that was assigned a goal of reducing waste across a campus. They quickly determined, through personal experience and with no data support, that a leading source of waste was extra paper usage at the copy machine. Just as quickly, they reset all copiers to a two-sided copying default mode and declared victory. In the absence of overall context (why are we doing this?) and a strong communications campaign that was supported by persuasive data, employees quickly learned how to reset the machines and grew resentful of a team that was forcing them to change their copier habits.

Lack of Authority

Another failure mode is the act of launching teams without providing them with proper authority to implement their ideas and recommendations. In an example from our files, a large manufacturing organization launched a cross-functional team that was asked to focus on reduction of pollutants in their wastewater. Through some data analysis, the team determined that a quick win with measurable impact would be to change the soap being used in all the soap dispensers across 15 plants. The new product was both environmentally friendly and less costly (20 percent less) than the offending product. The team launched a pilot program at the headquarters site. Employees noticed a change in the color of the soap and began to ask questions. No complaints about effectiveness were received, just general questions about why the soap was changed. The plant manager decided he didn't want to deal with the employee noise and ordered a return to the old soap. Obviously, a better communications campaign would have helped, but in the absence of any formal authority, the team was not even able to get a meeting with the plant manager to discuss the case for the change.

Insufficient Data and Tools for Analyzing the Data

Another chronic failure mode for teams is the inability to gather sufficient data or to perform appropriate data analysis to make good decisions. Early in the life of most teams, they enter into a stage of brainstorming to determine the most significant areas where their efforts can have the greatest impact on the overall sustainability targets. In the absence of good data with which to judge alternatives or to measure potential impacts, teams will rely on gut instinct, emotion, or the person who most desires to exert his or her will on the majority. In the world of green project teams, too many teams launch well-meaning efforts based on emotion or passion for the cause (Save the whales!) and have very little opportunity for impact.

Weak Leadership

Of all the reasons that green project teams fail, the most significant failure mode is weak leadership of the team. In the absence of strong leadership, teams will waste many days in the early stages of team development—the forming and storming stages—and often get stuck and disband. They are unable to drive conversations to conclusions, unable to agree on meaningful goals, and have great difficulty in building and committing to an action plan. Even if they can get to goals and create an action plan, such teams rarely execute in time to make a difference. Too many green project teams celebrate the birthday of the formation of the team without ever achieving meaningful impact. All this occurs primarily because of the absence of an effective team leader.

Eliminating the Potential Failure Modes

It is for these reasons that self-directed teams tend to fail in general and why they fail in particular when focused on sustainability-related projects. We have watched while well-meaning and enthusiastic green teams work feverishly on projects that may boost employee feelings about caring for the environment but actually have very little (or negative) impact on the environment that they believe they are changing. Figure 5-1 presents some examples of these misguided team efforts.

The tragedy here is that all this creative energy could have been channeled into projects and teams that actually have an impact on the causes that team members care about. Successful teams working on

"Empowered" Teams That Failed to Make an Impact	Why They Failed
"Turn the lights out" poster campaigns	Lack of meaningful data to support the new behavior
Annual "bike to work" day	Lack of leadership, no ability to measure impact, lack of goal clarity
Local "save the polar bear" campaigns	Decisions based on "gut" instinct, no data, lack of goal clarity; lack of affordable, practical solutions
Site-based "change the light bulbs" campaigns	No authority to enforce solution over time

Figure 5-1 Examples of misguided team efforts.

sustainability goals have certain success factors in common. They have clear and measurable goals, committed sponsorship, data and tools from which to draw fact-based conclusions, effective team leadership, and the authority to make decisions and drive implementation. Fortunately, for all aspiring green project teams, we can look to our Six Sigma methodology for proven practices in how to enable these success factors.

The Six Sigma Methodology for Driving Team Success

As we reviewed the Six Sigma body of knowledge for practices that help to ensure team success, we identified three specific practices that are most critical to team success:

1. Development of detailed, one-page team charters
2. Establishment of a supportive leadership structure
3. Adoption of a consistent team problem-solving model known as define, measure, analyze, improve, and control (DMAIC)

In the pages that follow we will detail these three practices.

Development of Detailed, One-Page Team Charters

Team-based improvement efforts are linked directly to the scorecard through leader-developed team charters. Team charters provide a tool through which projects and objectives are both developed and then communicated to team members and all stakeholders connected to the project. Charters help to clarify and communicate what the prioritized projects are and what team resources are assigned to that project.

Charters help the team to clarify what specific outcomes are required from the project. Through the chartering process, teams develop a specific, defined scope for the project and determine the completion date, as well as the milestones that will drive toward that completion date. Most important, the charter represents the document through which the team commits its intentions to its sponsor and the sponsor communicates his or her alignment with the efforts of the team. The charter is an important tool to help avoid the team failure modes of lack of goal clarity, projects that are too narrowly focused, lack of clear authority, and lack of alignment with an executive sponsor.

In order to promote proficiency and repeatability in the chartering process, the Six Sigma community has developed a consistent format for the one-page charter document. Every charter has six basic elements: the *business case*, which states the purpose of the project; the *opportunity statement*, which defines the desired business impact; the *goal statement*, which establishes success criteria for the project; the *project scope*, which defines the boundaries of the project; the *high-level project plan*; and finally, a *list of team members*.

Given the importance of the charter, it is useful to understand each element and the questions that should be asked and answered in the completion of each element (Figure 5-2).

Business Case

The *business case* describes the business benefit for undertaking a project. The business case addresses the following questions: Does this project align with other business initiatives? What is the focus for the project team? What impacts will this project have on other business units and employees? What benefits will be derived from this project? Does it make strategic sense to address this problem? Does the project support or leverage other sustainability initiatives?

Opportunity Statement

The *opportunity statement* describes the "why" of undertaking the green initiative. The opportunity statement should address the following

Business Case	**Opportunity Statement**
This project supports the corporate goal of becoming the number 2 global financial services company by increasing customer retention and satisfaction.	An opportunity to reduce customer defection (27% of applicants) and reduce cost may be achieved by improving our loan and lease processes. The loan and lease processes currently have an average cycle time of 9.2 days which is worse than our customer requirement of 8 days and our application processing cost exceeds the application fee by 18%. Customer defections represent a revenue loss of $2,500,000 per year and a cost of $165,000 for partial application processing. Current Sigma is 1.6.
Goal Statement	**Project Scope**
Reduce average load and lease cycle time to 6 days by Oct. 1. Improve Sigma to 3.0 by Oct. 1. Reduce processing cost by 20% by the end of the year.	Loan and Lease Processes—Begins with a call from the customer and ends with the acceptance or rejection letter sent to the customer.

Project Plan			**Team Selection**	
Activity	Start	End	Albert Anderson	Champion
Define	5/1	5/15	Carrie Carson	Master Black Belt
Measure	5/10	6/10	Barry Bethel	Black Belt
Analyze	6/5	7/20	Denise Davidson	Customer Service
Improve	7/15	8/15	Eric Edwards	Sales Representative
Control	8/15	9/15	Frank Fischer	Loan Department
Track Benefits	10/15			

Figure 5-2 Example team charter.

questions: What is wrong or not working? When and where do the problems occur? How extensive is the problem? What is the impact on our customers? What is the impact on our business? What is the impact on our employees?

Goal Statement

The *goal statement* defines the objective of the project in measurable terms. The following questions should be answered: What is this green team seeking to accomplish? How will the team's success be measured? What specific parameters will be measured? What are the tangible, "hard" deliverables/results (e.g., reduced CO_2 emissions, reduced cost, reduced drive times, etc.)? What are the intangible, "soft" deliverables/results? What is the timetable for delivery of results?

Project Scope

The *project scope* defines the boundaries of the opportunity (Figures 5-2). The following questions should be answered: What are the boundaries (the starting and ending steps) of the initiative? What parts of the business are included? What parts of the business are not included? What, if anything, is outside the team's boundaries (e.g., "This project will focus on office space in Chicago only")? In good project management, the project scope and the project schedule (Figure 5-3) are managed together since changes to one generally cause changes to the other.

Project Plan
Purpose: To document the major milestones and timing

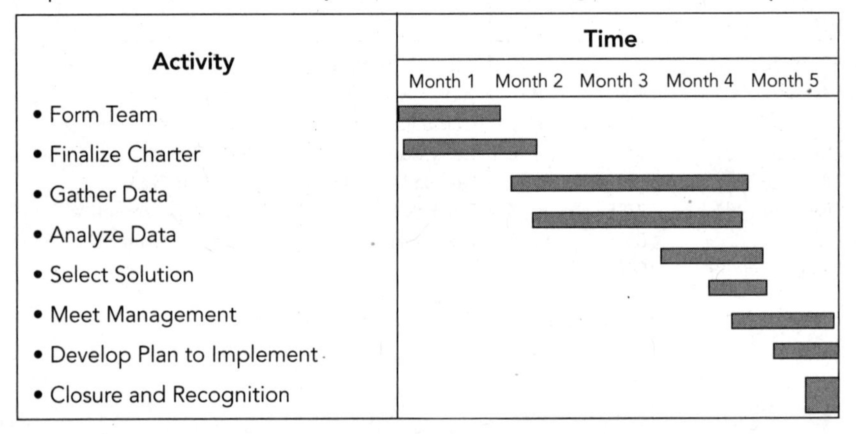

Figure 5-3 The project plan.

Team Selection

The purpose of *team selection* is to identify the team members and other individuals who will be required to support the team and to delineate the responsibilities of each individual named. Questions to be answered include

- ▲ Who is the project champion? What are his or her responsibilities to the team?
- ▲ How will project team members coordinate their efforts?
- ▲ Who is the team leader? What are the team leader's responsibilities?
- ▲ How will the team report its findings and how often?
- ▲ What types of team members are needed? At what stage will they be needed?

Team Charter Evaluation

Once the team has completed a draft of the team charter, team members should evaluate the charter to ensure its effectiveness. One methodology for evaluation is called *SMART*.

This acronym is a checklist to ensure that the charter is effective, thorough, and actionable. The components are defined as follows:

- *Specific:* Does it address a real business problem?
- *Measurable:* Are we able to measure the problem, establish a baseline, and set targets for improvement?
- *Attainable:* Is the goal achievable? Is the project completion date realistic?
- *Relevant:* Does it relate to a business objective?
- *Time bound:* Have we set a date for completion?

Establishing a Supporting Leadership Structure

Once the charter is developed, it is essential that a leadership structure be in place that will enable team success. This is another area where we can draw lessons from the Six Sigma body of knowledge. Through experience, the Six Sigma community has evolved a three-tiered leadership structure consisting of three specific roles. The roles are sponsors, champions, and team leaders.

Sponsors

As we noted earlier, the myth of empowerment is that empowered, self-directed teams will be motivated and creative and will do good things because we have freed them from the shackles of management oversight and structure. The fact is that, especially with regard to green project teams, most self-directed, empowered teams fail to make an impact on the organization's sustainability objectives. High-performing teams require strong guidance and the visibly engaged support of a senior leader. The knowledge gained from other Six Sigma team successes is that the senior leader should be carefully selected and designated the formal role of team sponsor. The role of the sponsor is as follows:

- Help to align the team with the overall sustainability objectives.
- Provide the team with the resources needed to achieve its objectives.

- Remove organizational barriers.
- Provide rewards and encouragement.
- Communicate the mission of the team, and engage others in support of that mission.
- Communicate and celebrate the success of the team.

Champions

Since sponsors are, by definition, senior leaders, it is unlikely that they are close to the problem or have the time to provide deep operational support to a team on a weekly basis. For this reason, the Six Sigma community created the role of champion to be played by a first-level manager who is likely to benefit from the project, who is close enough to the work of the team to understand if they are on track, and who can provide guidance and coaching as required. Typical roles of the champion would be as follows:

- Assist, encourage, and gain support for black belt candidates.
- Identify and rally key players.
- Remove roadblocks.
- Muster consensus.
- Ensure that the Six Sigma methodology is followed appropriately.
- Review projects regularly, stop a project if necessary, and ensure advancement of the improvement projects.
- Challenge the status quo.
- Provide genuine and committed leadership.

The champion fulfills this role through a specific set of tasks:

- Conduct weekly reviews with project teams.
- Ask the right questions throughout the DMAIC process.
- Ensure the financial returns of the projects.
- Manage and accelerate change associated with implementing improvements.

With a sponsor designated and an active champion in place, the green project teams have the appropriate leadership support to keep them aligned with the overall organization, focused on the right objectives, and moving to complete the project on an appropriate time line.

Team Leader

One final level of leadership required to ensure the success of a team is the individual who will be an active team member and drive the day-to-day activities of the team. The team leader's responsibilities are as follows:

- Facilitating team meetings
- Proactively engaging all team members
- Encouraging balanced dialogue across the team
- Establishing the project plan
- Holding team members accountable for delivering on commitments
- Reporting progress to the champion
- Soliciting support from the sponsor and champion as required

Adoption of a Consistent Team Problem-Solving Model Known as DMAIC

The DMAIC problem-solving methodology and the associated tools and training to support the methodology have evolved over the past 20 years to become a set of powerful, robust, and widely adopted practices. The methodology was specifically developed to help teams get to root-cause problem solving more efficiently and with greater consistency and repeatability across teams. Since there are entire books and courses that teach the methodology, our purpose here is to provide an overview to help the reader gain an appreciation for how the methodology can be applied in the green team arena and encourage team members to learn the methodology and supporting tools.

The DMAIC problem-solving methodology (Figure 5-4) was developed to help teams answer five key questions with regard to any problem or opportunity:

- What is important?
- How are we doing?
- What is wrong?
- How can we improve?
- How do we guarantee sustainable gain?

Each question triggers one of the five phases to form the sequence define, measure, analyze, improve, and control. We will describe each phase.

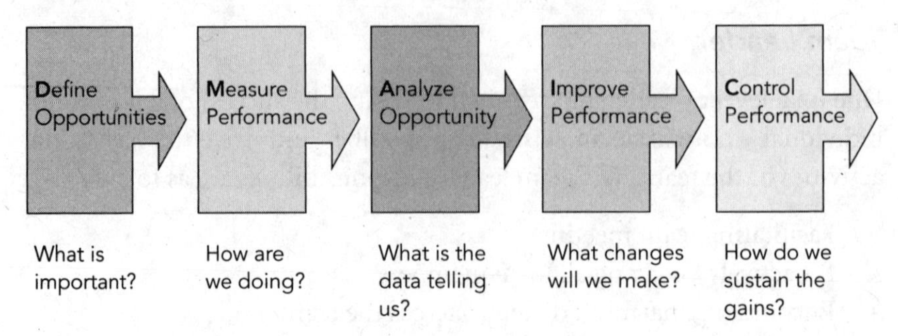

Figure 5-4 DMAIC improvement methodology.

Define

The purpose of the define phase is to identify and/or validate the project opportunity, develop the process that will drive the green initiative, define critical stakeholder requirements, and prepare team members to act as an effective project team. Ideally, teams will work through the define step in a single eight-hour workshop. This focused session has the effect of pulling the team together around a common understanding of the green problem that they are trying to solve and the goals and objectives that they share. During the session, team members will hear from their sponsor regarding the importance of their project and how the project aligns with overall sustainability goals of the organization. The define session is most effective when the champion coleads the activities with the team lead. Key activities of the define phase (Figure 5-5) include the following:

- Validate/identify the green improvement opportunity.
- Validate/develop the team charter.
- Identify and map processes.
- Identify quick wins, and refine the work process.
- Gather expectations of various stakeholders and convert those expectations into critical project requirements.
- Develop team guidelines and ground rules.

Using our Apex case study from Chapter 1 as an example, let's assume that a team has been chartered by the chief financial officer (CFO) to determine ways to reduce overall energy consumption in the Apex data centers. In the initial define workshop, the team spent an hour on the phone

Objectives	Activities	Tools	Deliverables
• Identify the improvement opportunity • Develop the current state process • Define critical shareholder requirements • Prepare to be an effective project team	• Create team • Develop team charter • Perform stakeholder analysis • Understand voice of the stakeholder • Document process map • Identify barriers within process • Perform value stream analysis	• Team charter • Stakeholder analysis • Flowchart • Value analysis	• Prioritized shareholder requirements • Current state process maps • Clear team charter • Quick wins

Figure 5-5 The define phase.

with the CFO (the official sponsor) and heard about the critical role that data centers play in the overall consumption equation at Apex. The CFO was clear that the team charter was to understand where energy was being wasted in delivering power to the servers and maintaining temperature, but the charter would not include delving into server and application operations. The CFO also was clear that it was critical that the team's work could not cause an interruption in operations or negatively impact center reliability. In that one-hour phone call, the team learned the business case, formulated the opportunity statement, captured the voice of a critical stakeholder, and began to understand critical requirements of the project. Working with their champion, team members documented what they had heard and moved into a working session to map the energy flows within the data center. They also mapped the work process through which energy consumption was currently being managed and monitored. During the mapping process, team members began to see where improvement opportunities might exist and planned their early quick wins. This activity helps to get the team excited about the potential for the project and motivated team members to set an aggressive work plan and agree on team norms. With its define workshop completed, the team was ready to move into the measure phase.

Measure

A key lesson learned from our review of why green project teams fail is that a critical failure mode for green project teams is taking action in the absence of meaningful data. Teams work from personal knowledge and gut instinct and take actions that may or may not have an impact. In order to avoid that failure mode, the Six Sigma community has formalized various methodologies for collecting relevant data and measuring process performance. These techniques have been documented as the measure phase in the DMAIC sequence.

In the measure phase, teams determine what they should measure and what techniques and tools they can use to conduct the measurement and data collection, and then they review methods for ensuring that their measurement process is valid and accurate. Once the measurement plan is in place, the measure phase continues as the measurement and data collection take place. Data collection continues until the team finds that it has a statistically valid sample size from which to conduct valid data analysis.

Key to the success of any green team is developing baseline performance data during the measurement phase. The baseline data inform that team about how the process is currently performing and whether or not improved performance can make a significant impact on the overall target. In the absence of baseline data, teams make what they think might be improvements but have no ability, after the fact, to prove that they have indeed improved performance, saved money, reduced consumption, or affected the overall carbon footprint. Typical activities during the measure phase (Figure 5-6) include the following:

- Determine process performance.
- Identify input, process, and output indicators.
- Develop operational definitions and a measurement plan.
- Plot and analyze data.
- Determine if special causes exist.
- Collect other baseline performance data.

Turning back to our Apex case example, the team chartered to improve data center energy consumption determined, in its measurement plan, that it was necessary to first develop an accurate method for measuring energy drawn into the data center on a nearly real-time basis and identified a low-cost package that included specific metering and software for collection and reporting. The team also developed an assessment tool to determine where leakage, or unnecessary consumption, within the data center was occurring. For example, team members observed missing floor panels, fans that were

Objectives	Activities	Tools
• Identify key measures to evaluate the success, meeting CCRs • Establish baseline performance for the processes the team is about to analyze	• Identify input, process, and output indicators • Develop operational definition and measurement plan • Plot and evaluate data • Determine if special cause exists • Determine performance level • Collect benchmark data	• Flowchart • Data check sheet • Benchmark data collection • Surveillance • Graph and charting selection guide **Deliverables** • Data collection plan • Baseline data set

Figure 5-6 The measure phase.

running continuously, and lighting at unnecessarily high levels. By placing meters at five data centers and assessing conditions across all five data centers, the team soon reached a point where it had a valid sample size and was in a position to begin to analyze the data to determine root causes of unnecessary consumption and prioritize improvement opportunities. The team was ready to move to the analyze phase.

Analyze

The purpose of the analyze phase is to provide teams with the techniques and tools they need to stratify and analyze the data collected during the measure phase in order to identify a specific problem (root cause) and create an easily understood problem statement. When teams reach a point in which they want to analyze available data, they are confronted with two potential failure modes. These failure modes are either a lack of relevant data or too much data and an inability to determine how to analyze those data in ways that will lead to relevant conclusions aligned with the problem the team is trying to solve. Teams typically follow a process of first creating a problem statement or hypothesis of what the problem is (e.g., "Lighting is the number one source of energy loss in this data center"). Then teams use data-stratification techniques, comparative analysis, and regression analysis to either prove or disprove the hypothesis. Teams will run through a number of hypothesis statements and the associated analysis until they can statistically prove that they have identified the sources of variation that are the most valid root causes of the problem. The list of activities and techniques employed by teams in the analyze phase (Figure 5-7) typically could include the following:

- Development of the problem statement
- Stratification of the data
- Comparative analysis of multiple data sets
- Performing sources-of-variation studies
- Analysis of failure modes and effects
- Regression analysis to determine the strongest correlations with the problem statement
- Identification of root causes
- Design of root-cause verification analysis
- Validation of root causes
- Design of experimental studies to statistically prove the root cause

In the example of our Apex case study, the team charged with reducing energy consumption in the data centers started first simply by analyzing utility billing data, but team members quickly learned that these data were not granular enough to determine the root causes of wasted energy within a specific data center. Such a determination led the team to develop very specific metering and monitoring strategies within each data center. Over time, the team developed data sets that showed energy drain from cooling units, lighting systems, specific servers, and missing floor panels. Data stratification and comparative analysis enabled the team to isolate the most likely root

<table>
<tr><th>Objectives</th><th>Activities</th><th>Tools</th></tr>
<tr><td rowspan="3">• Analyze the opportunity to identify a specific problem defined by an easily understood problem statement
• Determine true sources of variation and potential failure modes that lead to shareholder dissatisfaction</td><td rowspan="3">• Analyze current state
• Develop problem statement
• Identify root causes
• Validate root causes
• Perform statistical analysis
• Evaluate benchmark data compared to current state
• Identify performance gaps</td><td>• Run charts
• Control charts
• Cause and effect diagrams
• Statistical tools</td></tr>
<tr><th>Deliverables</th></tr>
<tr><td>• Source of variation study
• Validated root causes
• Problem statement
• Potential solutions</td></tr>
</table>

Figure 5-7 The analyze phase.

causes by comparing data sets across centers. Designed experiments enabled the team to conduct "what if" scenarios employing a variety of solutions.

It should be noted here that not all teams find it necessary to go deep into statistical analysis. Very often simple stratification techniques such as Pareto analysis (aka the 80/20 rule) enable teams to identify the 20 percent of root causes that drive 80 percent of the problem. The tip here is not to avoid analysis because of lack of statistical skills but rather to use the tools that are available in basic Excel and reach out to a coach or the champion if those tools don't lead to obvious root causes.

Once teams have identified the root causes of their problem, they are at the point in which they are anxious to begin working toward solutions. The improve phase is organized to provide teams with a methodology to sort through potential solutions and determine the best-fit solutions.

Improve

The purpose of the improve phase is to enable teams to identify, evaluate, and select the right improvement solutions and then to develop a change-management approach to assist the organization in adapting to the changes introduced through solution implementation.

As we reviewed green teams that were unsuccessful, we discovered a fairly common failure mode. Many teams were successful in their data collection and were able to present compelling analysis but fell short of actually implementing a solution because they had proposed a solution that was simply not practical or economically and culturally feasible. For example, a team can prove statistically that shifting 50 percent of the employee population to cycling to work as a transportation mode is environmentally beneficial. Economically, though, it is not practical to buy each employee a bicycle, and culturally, this solution is a tough sell to a major segment of the population.

In order to help avoid this particular phenomenon, the Six Sigma community has designed the improve phase so that it lays out specific techniques for finding the most creative and feasible solutions, building strong business cases, developing effective change-management strategies, and developing tight implementation plans. Teams typically will work through a series of workshops in which they begin by using various brainstorming activities, informed by research and subject-matter experts, to develop creative alternatives to solving their problem. Teams then work through various process designs to develop the process improvements that will enable the solution to be permanently hard wired into the workflow and the organizational processes. Once the solution is designed into the process, teams work through the financial analysis to demonstrate that the overall cost of implementation is far less than the economic gains resulting from the improvements. The result is a business case that provides appropriate rationale for the investment. With the business case in place, the team then goes to work designing a change-management strategy. We will discuss a change-management framework later in this chapter.

The point here is that even the best-designed solutions will fail unless the targets of the change have bought into the change and see a clear benefit to themselves personally. Designing the change-management plan therefore requires developing a clear vision and leadership strategy, analyzing the needs and concerns of various stakeholders, and developing a thoughtful communications strategy.

Once teams have developed their change-management strategy, they move into the final activity in the improve phase, which is the development of an overall implementation plan. In this activity, the team determines all the required tasks to implement a solution, develops a sequence for those tasks, estimates completion dates for each task, and assigns ownership for each task and for the overall implementation.

The typical sequence of activities during the improve phase (Figure 5-8) is as follows:

- Generate solution ideas.
- Determine solution impacts and benefits.
- Evaluate and select solutions.
- Develop the process map and high-level plan.
- Develop financial analysis and the business case.
- Develop and present the solution storyboard.
- Develop the change-management plan.
- Communicate the solution to all stakeholders.

The Apex case study provides a simple example of a team moving through the improve phase. The team charged with reducing energy consumption in the data centers determined that a key element of the solution was implementation of an energy monitoring and control system

Objectives	Activities	Tools
• Identify, evaluate, and select the right improvement solutions • Assist the organization in adapting to the changes introduced through solution implementation	• Brainstorm possible solutions • Perform a cost/benefit analysis • Reach agreement on solution • Perform pilot and review results • Make business case • Launch Blitz teams • Design and execute implementation plan	• Brainstorming • FMEA analysis • Process simulation • Staff feedback • Surveillance • Implementation planning **Deliverables** • Ideal process design • Business case approved • Implementation plan • Blitz team charters

Figure 5-8 The improve phase.

across the data centers. Working from the process maps developed in the define phase, the team determined how the monitoring and control function would be managed and then designed those activities into the work processes. Data from the measure and analyze phases helped the team determine where the monitoring and control units would have the greatest impact. Armed with data from the measure phase, the team was able to put together a financial analysis that proved full payback within 12 months. Working with their sponsor, team members developed a clear vision and case for change that would engage the various stakeholders. Then the team worked with the stakeholders to develop an implementation plan that was practical and achievable.

Confident that it has designed the best-fit solution, supported by an approved business case, a solid change-management strategy, and a tight implementation plan, the team is in a position to move to the control phase.

Control

The purpose of the control phase is to help teams understand the importance of planning and executing against the plan and to determine the approach to be taken to ensure achievement of the targeted results. The control phase also helps teams to understand how to disseminate lessons learned, to identify replication and standardization opportunities/processes, and to develop related plans. Most important, the control phase forces teams to think through strategies so that identified benefits and financial impacts actually will be realized when the solution is fully implemented and institutionalized. It also will ensure that the solution will deliver results over a long period of time.

When we reviewed the work of green project teams that have failed, one of the most disturbing failure modes we saw is failure to achieve permanent implementation of the solutions. Often solutions are piloted successfully, announced with great fanfare, and then allowed to languish. General resistance to change, lack of leadership focus, and failure to permanently wire the solution into organizational operating processes are all reasons that this failure mode occurs. The control phase is designed to force teams to understand how this failure mode could occur with their solution and to design strategies that ensure that this failure will not occur in their implementation.

Often this analysis of potential implementation failure modes occurs by conducting controlled pilot programs that enable teams to study what works and what might go wrong. In addition, teams will work with the data sets developed during the measure and analyze phases to determine key process performance measures that, when monitored closely, will inform leaders as to the ongoing performance of the solution and enable leaders and teams to monitor and improve the solution in order to guarantee the desired results.

Typical activities that occur during the control phase (Figure 5-9) are as follows:

- Develop the pilot plan.
- Conduct and monitor the pilot.
- Verify reduction in root causes resulting from the solution.
- Identify whether additional solutions are necessary to achieve goal.
- Identify and develop replication and standardization opportunities.
- Integrate and manage solutions into the daily work processes.
- Integrate lessons learned.
- Identify the team's next steps and plans for remaining opportunities.

We can demonstrate these activities in the Apex case study. The team charged with the responsibility of reducing energy consumption in the data center had determined in the control phase that a part of the solution would be the installation of a monitoring and control system across all the data

<table>
<tr><th>Objectives</th><th>Activities</th><th>Tools</th></tr>
<tr><td rowspan="3">• Understand the importance of execution against the plan
• Assure targeted results
• Disseminate lessons learned
• Prevent reversion to current state</td><td rowspan="3">• Determine approach to assure targeted results
• Track metrics that will show if ideal process is in control
• Review progress reports regularly and adjust as needed to support adoption of new process</td><td>• Control charts
• Statistical process control
• Leadership and change management</td></tr>
<tr><th>Deliverables</th></tr>
<tr><td>• Process control plan
• Ongoing monitor and reporting plan
• Replication opportunities</td></tr>
</table>

Figure 5-9 The control phase.

centers. The team determined that it would be appropriate to pilot the system in a single system. During the pilot, team members were able to test and adjust various sensor types. They learned that certain work routines had to be modified in order to enable near-real-time monitoring, and key employees had to be trained in order for them to understand what actions to take when presented with various readings from the monitoring system. Most important, team members learned that the engineers really didn't see this new activity as part of their job and viewed the monitoring as "big brother" watching them. The pilot did, however, prove out the savings anticipated in the business case. As a result of this knowledge, the team improved the work routines, added new detail to the communications plan, and recommended a share-in-the-gains incentive system that would encourage the engineers to actively embrace the new system.

As a result of completing activities similar to those just described, teams are able to carefully design control plans that ensure lasting gains.

Blitz Teams

A key challenge in implementing green solutions is the ability to spread a solution across multiple geographies or multiple sites. In order to address this challenge, the Six Sigma community has developed the concept of *blitz teams* to aide the core Six Sigma team in the implementation phase. Blitz teams require the same leadership structure as the core team and are formally chartered, just as the core team was chartered. But blitz teams don't work through the measure, analyze, and improve phases. Their responsibility is to accept the analysis and solution developed by the core team and basically drive implementation, following the implementation and control plan designed by the core team. Blitz teams are a formal method to drive implementation rapidly across multiple locations or departments.

Summary of DMAIC

In summary, the DMAIC problem-solving methodology, as well as the associated tools and training to support the methodology, is a powerful, robust, and widely adopted set of practices designed to improve the success rate of problem-solving teams. The methodology was developed specifically to help teams get to root-cause problem solving more efficiently and with greater consistency and repeatability across teams. This overview was developed to help the reader gain an appreciation for how the methodology can be applied in the green project team arena and encourage team members to learn the methodology and supporting tools.

While the DMAIC methodology provides teams with the process and tools required, that methodology is not sufficient to ensure that the solutions developed will achieve any level of organizational acceptance and adoption. Throughout a sustainability initiative, the leadership team must implement solid change-management strategies to ensure that the team remains committed, the overall organization understands and supports the sustainability objectives, and the organization therefore is ready to support adoption of the green project team's solutions.

Change Management

In order to fully realize the impact of the efforts of green project teams, leaders must understand and consider the challenges of driving large-scale change across an organization. There are fully 65 books on Amazon.com with *change management* in their titles. Thus our aim here is not to summarize the body of knowledge focused on change management but rather to provide some practical tools and frameworks that we have found to be particularly useful in our work with individual teams and whole organizations.

Change has been the focus of business research and forecasting for decades, and yet current opportunities to create a truly sustainable workplace challenge us with the kind of changes for which few can claim to be prepared. Future technology promises radical and indeed inconceivable changes in how we work with and relate to one another. At the same time, our once-dependable resources—natural, human, and economic—appear irretrievably imperiled. Amid the cavalcade of rapid and unpredictable business changes, one thing is constant—*resistance to change.*

If there was ever a critical opportunity to examine how business leaders initiate, respond to, and manage change, it could be in the methods we employ to drive sustainability initiatives. In the following pages we will offer our view of practical, fundamental concepts in effective change management that should be considered when designing both large-scale and project-level green initiatives, including transformational processes, innovative organizational design, and a call for individual leadership that invites and rewards change.

Do You Change or Just Think about Change?

Managing business change is a popular topic for theoreticians. Thousands of books have been written on organizational change and adaptive behavior. Our familiarity with change is often not because of the varied events we have experienced but because of the catalog of ideas we have been exposed to.

From inside an organization, it can seem as though change always originates someplace else—as a management initiative, trend, or just a passing whim. While it might appear that compelling new ideas instigate change in an organization, ideas themselves never drive change. People do. People motivate, initiate, and achieve change, or, when poorly led, they don't. In an era where fast and far-reaching change is required, it is practical for leaders hoping to drive a sustainability initiative to ask themselves if they are the kind of person who can easily change and thereby lead others to do the same. We have learned, as we review failed sustainability initiatives, that business leaders cannot fake a commitment to green. It takes real, committed leadership and an understanding of the nature of change management to drive the level of organizational, cultural change required to drive sustainability across an organization.

Myriad Theories, Universal Formula

Analysts advance their own language and ideologies for change, but the underlying formula common to most of them can be expressed simply as:

$$Q \times A = E$$

Translated, this formula mean that, the *quality* Q of the solution times the *adoption rate* A determines the *effectiveness* E level of the result (Figure 5-10).

This formula is elegantly self-evident but runs counter to the way change is often approached. Usually, more time and focus are spent on the technical quality of a solution than the psychology of adoption, and as a result, acceptance is poor. Real change always occurs in response to an internal event—acceptance—and not an external stimulus. Achieving higher effectiveness therefore is a function of helping more people adopt change. Adoption is the most crucial factor of the equation and the subject we explore further. Applying this formula to the challenges encountered when driving sustainability initiatives, we find no shortage of elegant solutions being advanced in books and articles and many creative ideas being generated by green project teams. Clearly, creative, high-quality solutions are being designed and proposed to address most sustainability challenges that we face today. The challenge is not in the generation of creative solutions but rather in driving the adoption of those solutions across entire organizations.

In order to address the challenge of driving organizational adoption, we look to our Six Sigma body of knowledge to understand how successful leaders have been driving the changes that often evolve from Six Sigma

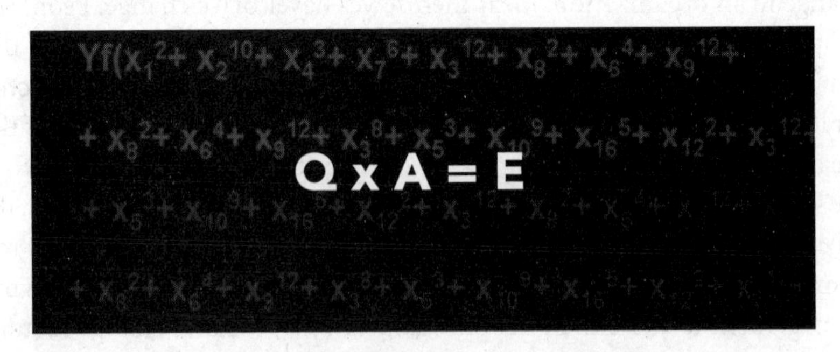

Figure 5-10 The effectiveness formula.

process-improvement initiatives. In order to drive the success of its Six Sigma initiatives, General Electric (GE) recognized that a change-management model was important. The GE Change Acceleration Process is a five-step approach to driving adoption of quality solutions within a group or organization (Figure 5-11):

1. Create a shared need.
2. Shape a vision.
3. Mobilize commitment.
4. Monitor progress.
5. Make the change last.

Each phase of the change-management process contains a specific set of actions that leaders should follow to ensure success of a sustainability initiative.

Create a Shared Need

In phase one, a small leadership team works together to develop its rationale and vision for what the team hopes to achieve in its sustainability initiative. To ensure that customers and other stakeholders will support the initiative, the leaders work to create the high-level business case for the changes they hope to drive. The business case considers the investments of time, money, and energy that must go into the initiative and weighs that against the estimated benefits of a successful initiative. With a clear vision and an understanding of the costs and benefits, the leaders begin to prepare the organization with early communication of the vision, rationale, and benefits, as well as their expectation of what the organizational response should be. In conjunction with these communications, leaders will sponsor an assessment process to determine the readiness of the organization to take on a sustainability initiative at this time.

Shape a Vision

In phase two, the small leadership team brings in leaders from across the organization and shares its vision, rationale, and business case for a sustainability initiative. In a workshop format, the extended leadership team is given the opportunity to dialogue about the pros and cons of the initiative

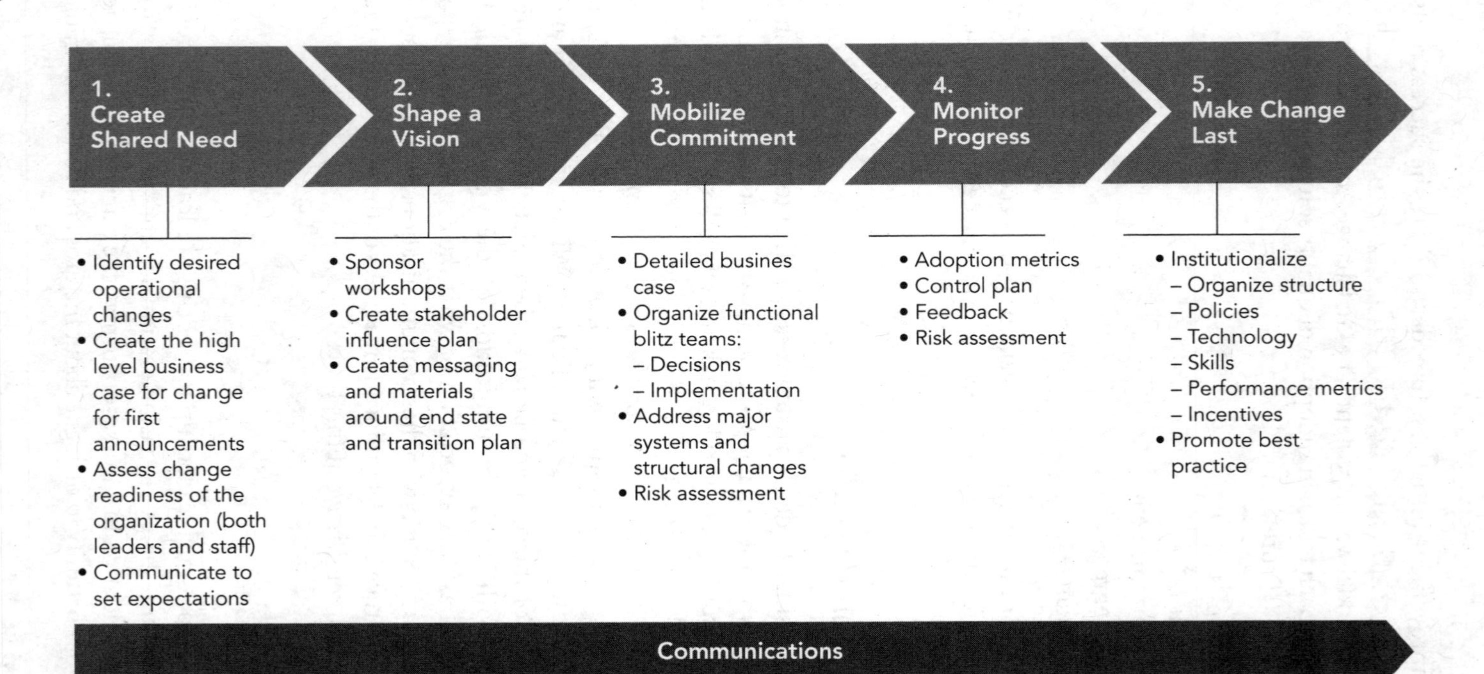

Figure 5-11 The change-management process.

and help shape the details of the initiative. This expanded team also works to create a stakeholder influence plan that delineates who the key stakeholders are, what their concerns might be, and strategies for engaging with those stakeholders to secure their support. With the vision, rationale, business case, and stakeholder influence plan in hand, the expanded leadership team creates a common set of messages and communications materials and begins the work of communicating the sustainability initiative across the organization.

Mobilize Commitment

At this point in the process, the leadership team hopefully has engaged the heads and hearts of the organization and therefore is positioned to engage their feet by getting them involved in meaningful action. Phase three is designed to engage teams in implementing green projects that will have a direct and meaningful impact on the initiative. These green project teams will engage in focused projects designed to identify the best opportunities for impact and to develop solutions to address those opportunities. Each project team will consider the major systems, key processes, and structures that must be improved to achieve impact. These improvements are supported by detailed business cases and a clear understanding of the risks associated with implementation. Change and improvement therefore are driven through these focused green project teams.

Monitor Progress

Phase four is designed to enable the leadership team to monitor and manage the pace and impact of the teams as they perform their projects. It is also used to monitor the acceptance of the organization as the teams drive for organization-wide adoption. A performance-measurement dashboard enables the leaders to monitor progress of the teams and adoption rates across the organization. A control plan provides strategies to ensure that each team is on track and that risks are being managed appropriately.

Make the Change Last

As projects are implemented and demonstrate positive environmental impacts, leaders will want to ensure that the gains they achieve will be sustainable over time. Phase five of the model is designed to help leaders drive institutionalization of the changes being implemented. In this phase, work processes are reengineered, organizational structures are modified, new policies are created, and supporting technologies are installed. Focused training enables key individuals to acquire relevant knowledge and to build required skills. Continued monitoring of the performance dashboard keeps the initiative on track, and incentives reward the appropriate new behaviors.

By carefully following these five phases, leaders will greatly improve the odds of success and ensure that the least amount of pain is inflicted on their organization. Stakeholders will embrace the sustainability initiative, and impact will be achieved.

Easy to understand, the process nonetheless is difficult to implement. The success rate of corporate change programs is said to be just over 30 percent. And while change initiatives are attempted by many, few executives have experience in managing large change campaigns effectively. To be fair, most leaders have dealt with the rise and fall of economic cycles, right sizing, and merger and acquisition integrations—all of which present difficulties of their own kind. Yet, for decades, most of the business of American organizations has been conducted in a stable and steady manner, consistent with past practices, insulated from interference, and free from the frightening imperative to change or fail.

As the inevitability of world-altering change approaches, it is worthwhile to examine what it requires to overcome resistance and lead change effectively. Resistance to change feels instinctual to most employees—and the instinct is fear. When fear is triggered, the benefits of change are easily overlooked or severely understated. Fear obscures reality, and the reality of change can be extremely beneficial on both organizational and personal levels.

In the same way that positive change can be institutionalized, so too is resistance.

Many presumptions about organizational behavior are actually examples of institutionalized resistance. Two issues, in point, are outsourcing noncore services such as energy management and building collaborative workspace to increase employee productivity while dramatically reducing the carbon footprint of the physical workplace. Fear of outsourcing can keep executives trapped in managing low-value, low-recognition functions. Resistance to collaborative workspace ignores the overwhelming personal preference for mobility.

Fear is also at work when we see organizational resistance to advanced, greener technologies and enforceable reduced space standards—both of which are reasonable and rational solutions to carbon-footprint reduction that, nonetheless, have relatively poor adoption and therefore minimal execution.

When we empower resistance, it discounts the organizational, strategic, and personal benefits of change. On the contrary, when we examine change in terms of personal benefit, acceptance skyrockets. All change is personal, and unless it is presented and perceived favorably on a personal basis, it will fail to achieve institutional adoption. Leaders need to stop selling change and start helping each person assess his or her personal cost of change.

Leaders of sustainability initiatives not only must inspire change but also must help each person realistically assess his or her own personal cost of change by asking two questions:

1. What do I give up?
2. What do I get?

When these two questions are answered, reality will overcome fear. This approach was first advocated by William Bridges in his seminal book, *Managing Transitions: Making the Most of Change.*[1] Bridges insightfully recognized that change occurs one person at a time. Leaders must give every individual in the organization an opportunity to investigate those questions and determine the outcome of his or her change equation. Moreover, the process begins with the leaders themselves.

The insularity we sometimes observe in organizations suggests that many leaders attempting to drive sustainability initiatives, while *informed* about change management, have not conducted their own personal change assessment. They resist a new and different role for themselves. For these executives to adopt a more strategic role in their organizations and become leaders of change, they must ask themselves the assessment questions and arrive at conclusions similar to these:

1. What do I give up? *A traditional, narrow, and passive role in operations*
2. What do I get? *Status and value as a serious part of a senior leadership team that creates vision and drives change*

The examples of gives and gets in the area of sustainability are strong illustrations of this problem and the need for leadership to achieve buy-in. Resource efficiency programs are almost always seen as a reduction in service levels. By asking employees to use less material, less energy, and less water, you are a priori asking them to give up an old way of doing business. By asking employees to care about planetary efficiency, you're running the risk that they will equate this with using a lower-quality supply. And, by definition, you're now asking them to factor new information into their decisions, information that they didn't used to have to worry about. If your reasons are not compelling, there's a small chance of success. At Yahoo!, for example, many employees commute south from San Francisco to their Yahoo! offices. As a commute-reduction program, Yahoo! started a charter van service to bring employees from various pickup locations in San Francisco. Yahoo! could have just asked employees to share rides or to use buses or trains. But Yahoo! leaders recognized the need for making this service more compelling than the personal-schedule freedom that employees would give up. In order to make the commute-share vanpool service more compelling, Yahoo! chartered comfortable vehicles that were equipped with wireless Internet access. In this way, employees could avoid the stress of an hour-or-more commute, ride in comfort, and if desired, work with laptops connected to the Internet. In just under two years, employee support led Yahoo! to expand from 3 to 18 vans serving 140 Yahoo! employees commuting to 4 offices from 22 different cities, saving employees $200,000 in annual fuel costs. This translates to 1.9 million commuter miles not driven and 751 tons of CO_2 reduced annually. In an emergency situation, a Guaranteed Ride Home program provides riders with taxi vouchers to address concerns about being without their own set of wheels.

Leaders who are on board with change can assemble and germinate engaged, energized green project teams that will have a significant impact on the organization's sustainability objectives and yield significant returns on investment. As we discussed earlier in this chapter, high-performance teams are the engines that drive successful change programs.

Teams are the primary accelerant of change initiatives. Regardless of how favorable people may feel about the personal benefits of change, without action teams, adoption easily loses momentum, and execution fails. A change that is unilaterally imposed is one that is more likely to be resisted. Major change programs require team activation and mobilization.

Team-driven models for organization-wide change begin with passionate sponsors who inspire middle managers to become champions. Champions mobilize people into cross-functional action teams to investigate a problem, offer input, evaluate a proposed solution, and answer the question, "How will this change make my work life better?"

The use of teams propagates change because every time a team is populated and its contributions are validated, champions are added to the cause. Through individual networks of influence, every champion pollinates dozens more supporters. Soon, whole groups of individuals are vested in an idea. The cascading team structure transfers ownership of an idea to the individual. The use of teams has been shown to triple the adoption rate of high-resistance changes (Figure 5-12).

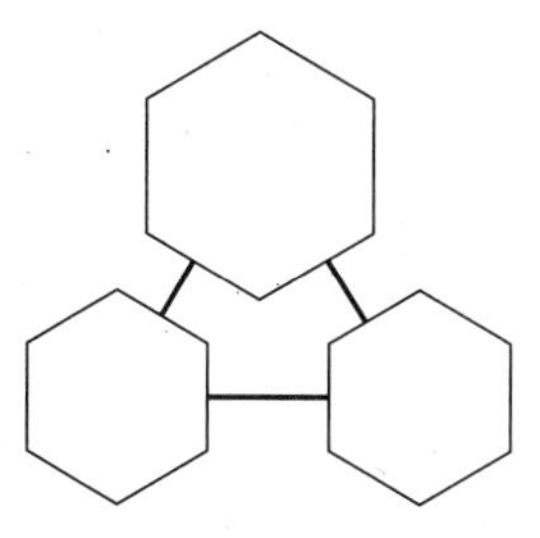

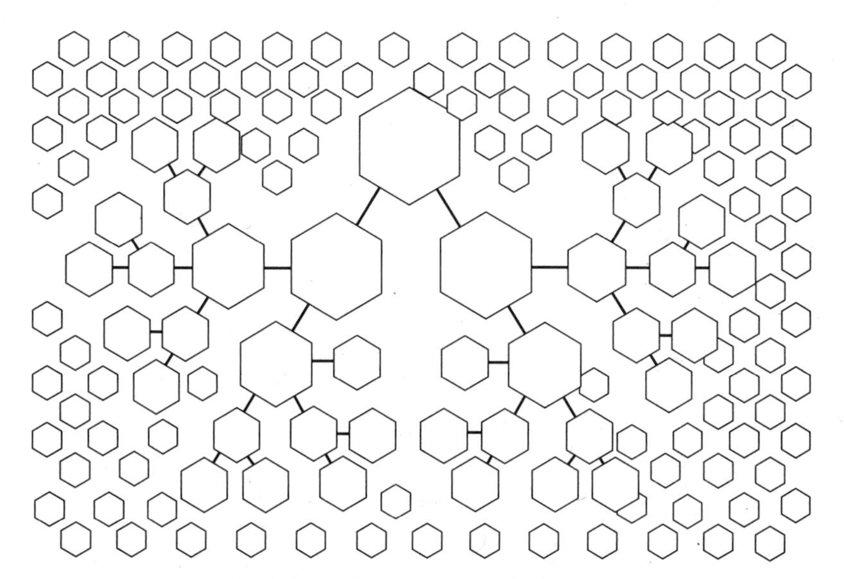

Figure 5-12 The multiplier effect of teams.

Organized for Change

Just as change programs designed around teams can achieve higher execution, organizations designed around change can far outperform conventional equivalents. Pyramid structures are inflexible and resistant to change, and yet they persist in many companies. The changing nature of today's work and workers better suits a structure that we referred to in Chapter 2 as the *shamrock organization.*

In that chapter we discussed the fluid alliances and short-term collaborations that have come to characterize workflow. We also described the three distinct groups of people who not only perform different kinds of work but also work for different reasons and rewards.

Core employees are the increasingly small and essential group of workers who own the organizational knowledge. Hard work and flexibility are expected of them; they are highly incentivized and well compensated. Yet, for these workers to remain committed and focused strategically, they have to be engaged in work they love to do. Tactical, day-to-day work is contracted out.

Strategic supplier alliances make up the second group. Not only is tactical work exported to this group, but so also is the risk of labor expansions and contractions. These workers excel at independently achieving day-to-day results so that the role of the core is to specify the results, not oversee the methods.

Subject-matter experts are the third component. These are people who engage in the science and research that can drive disruptive change. They can have values that are not necessarily aligned with those of your organization, but they need to be treated seriously so that their contributions remain valuable.

Although the formal structure of an organization may not reflect it, this is already the way most organizations function today: three types of workers collaborating by project and objective to accomplish goals. The key to success is in execution, and that depends on core workers having the time to anticipate, plan, and organize for strategic change.

When a shamrock organization is seamlessly aligned with its customer, it becomes a four-leaf clover with the customer as a fully integrated component of a dynamic enterprise. Boundaries are permeable, and people perform without a sense of whether they are inside or outside the organization.

This foundational shift in the idea of work and workplace is real today, proving that the change we might fear as "the end of life as we know it" actually occurs incrementally, every day, and is often undetected. Until we look around us with fresh eyes, we may not notice how much things have changed.

It is necessary to think about this type of an organization with a highly flexible, highly mobile workforce because a smaller, highly committed force of core workers adapts much more quickly to new practices and processes. A more flexible workforce requires much less permanent workspace and therefore enables significant reduction in the carbon footprint inherent in large office buildings.

Leading Change

While much is made of a leader's job to set expectations and drive agendas, open-mindedness is a virtue and skill during times of change. A strong leader is alert and receptive to change as it occurs. An agile leader listens and responds to feedback. A practical leader creates a flexible organization designed to respect and support changing needs, roles, and work relationships. An innovative leader no longer fears change but empowers teams of people to arrive at the best solutions, widest adoption, and highest execution by themselves. Leaders of change start by changing how they lead.

We can draw from our Apex case study for an example of leaders using teams to drive large-scale change. As part of an overall initiative to reduce costs, the Apex leadership team made a decision to relocate its corporate headquarters two miles from its current site. The current site was very costly to maintain and an energy hog. But it was located on a site where, if repurposed, would have high market value. Sale of the site would generate significant cash while enabling Apex to move to a new, energy-efficient, flexible building designed for a more productive, agile, collaborative workforce. The business case for both the investment and sustainability objectives was solid. The challenge, therefore, was overcoming employee resistance to giving up offices and working in an open, collaborative environment.

As ground was about to be broken on a headquarters building for same-city relocation, a workforce forecast indicated that the building would be full within six months of completion. Investment in a second building was unfeasible. A mobility program offered the best remediation, but resistance was high. Information technology (IT) argued that it would take two years to implement. Human resources (HR) estimated 18 months to change policy. Workers across the board wanted to keep their desks. A cascading team-based campaign was created with cross-functional groups charged with identifying, investigating, and selecting component solutions. Implementation times were a fraction of what had been feared. The headquarters move and mobility program were accomplished simultaneously under budget, on schedule, and with high employee engagement (Figure 5-13).

Six Sigma Integrated Team Framework as a Model for Driving Transformational Change

When all the components of the Six Sigma team implementation model are working together, the result is an integrated methodology for driving green project teams to successful results (Figure 5-14).

While the Six Sigma integrated team framework is a useful tool for driving success for teams such as the data center energy-management team, that same framework can be used as a model for driving wide-scale transformational change across an organization. To show how this strategy can be deployed to an organization-wide sustainability initiative, we will walk through the process of developing executive commitment and focus, engaging champions within the organizations, launching green project teams, and finally, achieving organization-wide adoption and bottom-line impact.

Sponsors' Launch

Applying the Six Sigma integrated framework for driving teams to this example, phase one would be the sponsors' launch. In this phase, the leadership team for the organization will come together with the support of some upfront analysis and a skilled facilitator in a two-day workshop designed to help these senior executives establish their vision and goals for a sustainability initiative. The first few hours of the workshop are focused on examining the various voices of the stakeholders that would be affected by a sustainability initiative. A simple question posed to each stakeholder

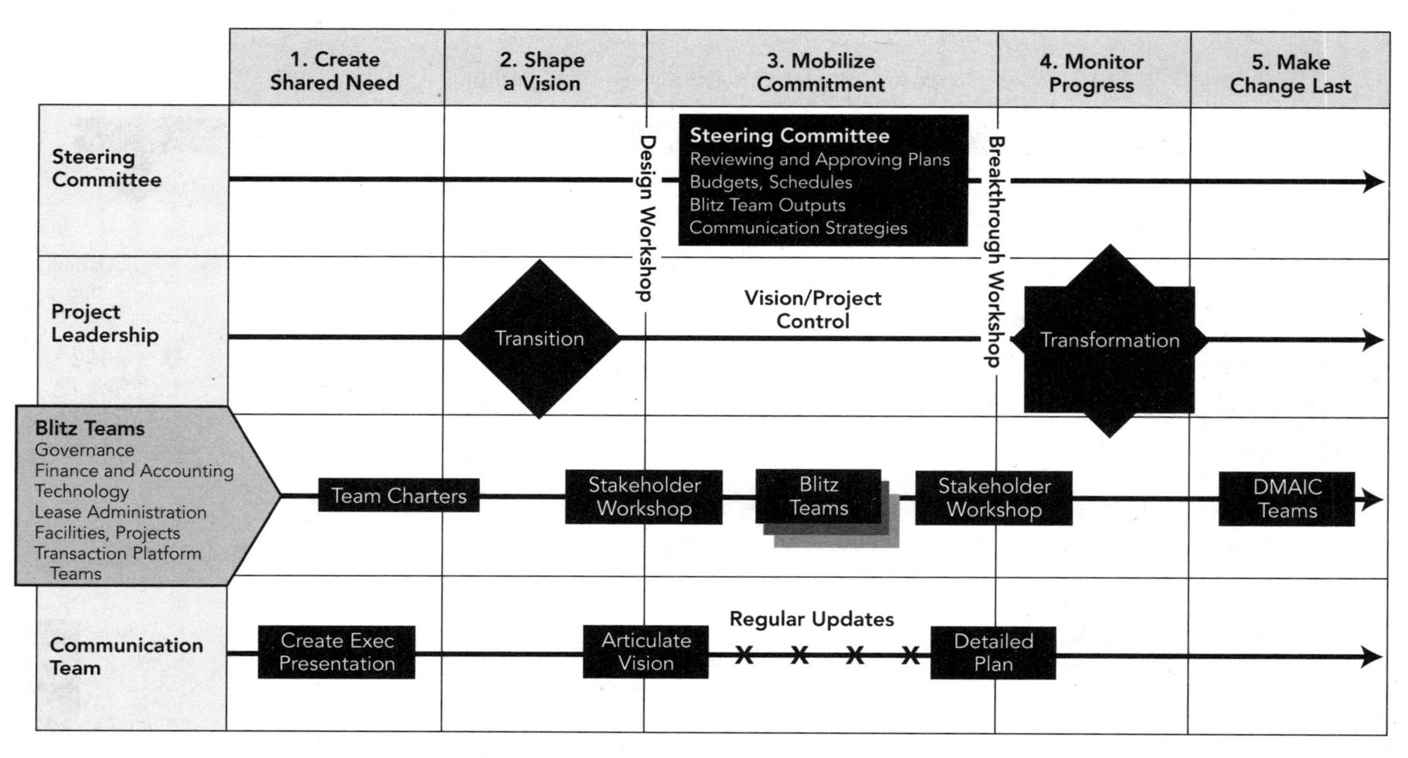

Figure 5-13 An integrated change plan.

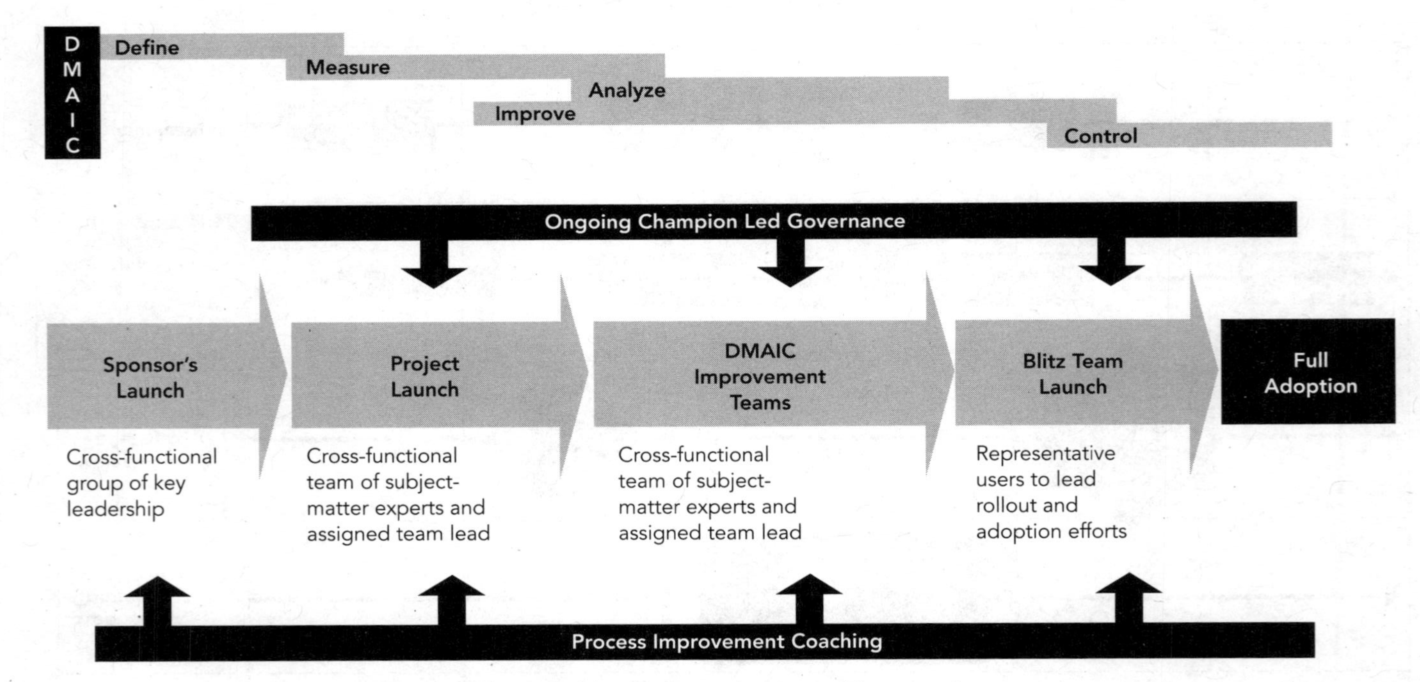

Figure 5-14 Six Sigma integrated framework.

group will bring out insight into what the stakeholders expect from a sustainability initiative. The leaders would brainstorm stakeholder responses to the following fill-in-the-blanks statement, "In order to meet my expectations, this sustainability initiative must. . . . "

Typical responses to this exercise could be as shown in Figure 5-15. The point is that in line with our Six Sigma methodology, customer requirements frame the issue and set the context for the sustainability initiative.

Once the team reaches common understanding of their stakeholder requirements, they move to an exercise designed to ensure that their sustainability initiative is well aligned with the overall strategic objectives of the organization. Typical outcomes of that activity would be as shown in Figure 5-16.

At this point, the sponsors have framed their sustainability initiative in a manner that ensures that their initiative is both supported by their stakeholders and positively contributing to the strategic objectives of the organization.

With stakeholder requirements and strategic objectives as their filters, the team is in a position to review the data that inform them as to where the potential opportunities for improvement might lie within the organization. This activity has the effect of educating the entire team so that all members

Stakeholder	Response
Customers	"In order to meet my expectations, your organization's sustainability initiative must provide me with a product that does not harm the environment and a product that costs less than your competition."
Employees	"In order to meet my expectations, your organization's sustainability initiative must provide me with opportunities to directly impact the effect that my work has on the environment and allow me to have a voice in how we improve our environmental impacts."
Shareholders	"In order to meet my expectations, your organization's sustainability initiative must deliver a positive return on investment, while reducing the overall organizational risks associated with negative environmental impacts."

Figure 5-15 Example brainstorming of stakeholder responses.

Strategic Objective	Alignment to Sustainability Initiative
Growth	Position our products to reach new segment of environmentally conscious consumers.
Cost Management	Reduce total product lifecycle costs. Reduce costs associated with our real estate portfolio.
Brand Image	Position our organization as a socially responsible member of the community.
Talent Management	Attract and retain employees that are proactive environmentalists.

Figure 5-16 Aligning strategic objectives with the sustainability initiative.

of the executive team have a common understanding of the situation and helps them to reach consensus as to which areas of opportunity they wish to attack within the initiative. The outcome is improved focus and shared commitment across the executive team.

This focus and commitment are expressed in a manner that illustrates the linkages of various improvement opportunities with the overall initiative. For illustration, we refer back to the sustainability transfer function introduced in Chapter 3. You will recall that the sustainability transfer function (Figure 5-17) established a clear linkage from the organization-wide superordinate goals ("big *Y*'s") of a large sustainability initiative to strategies and projects that ultimately would affect that superordinate goal ("big *Y*").

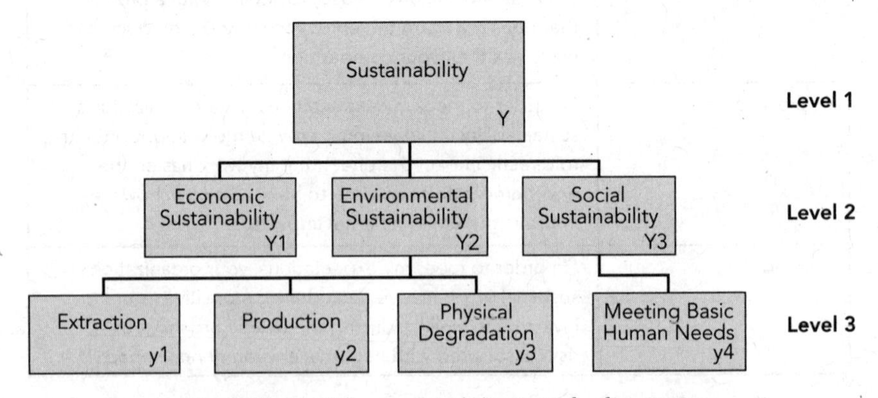

Figure 5-17 The sustainability transfer function.

Project	Champion	Objective
Landfill Diversion	Director of Supply Chain	Reduce amount of waste being trucked to landfills by 20%.
Procurement Practices	Sourcing Manager	Increase recycled content in our packaging by 40%.
Product Use Behavior	Director of Marketing	Increase consumer awareness of appropriate product disposal methods by 50%.
Transportation Distances	Director of Logistics	Reduce miles between distribution points and retailers by 20%.

Figure 5-18 Identifying projects, leaders, and objectives.

What flows from this activity are team discussions to determine what the first set of projects should be and who the sponsoring leader ought to be. Those leaders then become known as *project champions* (Figure 5-18).

As a result of this final sponsors' activity, the senior leadership team has set clear objectives, agreed on their improvement opportunities, established the projects that they wish to sponsor, and identified the leaders in their organization who are going to be charged with driving these projects. The initiative is ready to move to phase two, the champions' launch.

Champions' Launch

The champions' launch is a workshop in which the senior leadership team engages the leaders who are going to drive their initiative and secures their endorsement and commitment to the sustainability initiative and the projects that they will be leading. The workshop begins with a presentation from the senior leadership team to the champions in which the senior leaders explain the rationale and vision for the sustainability initiative. The senior leaders use the materials developed during the sponsor's workshop to communicate the rationale and vision and provide the champions with an example of how the champions will use the same materials to cascade the communications throughout the organization.

Having achieved the commitment of the champions to lead the sustainability initiative, the workshop transitions to an education mode.

During this segment of the workshop, champions learn the DMAIC model that we illustrated earlier in this chapter. The emphasis is on a process, tips, and tools that enable the champions to become effective coaches to the teams they must lead through the DMAIC analysis and improvement process. The final segment of the workshop is a set of activities in which the champions refine their project charters, identify their team members, and develop specific coaching strategies designed to move their teams effectively through the DMAIC process. At this point, the champions are ready to launch their green project teams.

Project Launch

The project launch follows much of the same process as the champions' launch. In this case, the champions deliver the vision and rationale for the sustainability initiative to the teams and work to secure the commitment of team members to the initiative and to their specific projects. Team members then are educated in the DMAIC methodology and begin the work of understanding how they will apply the DMAIC methodology to their projects. In the second day of the workshop, team members begin the activities that are delineated in the define phase of the model. They refine their project charters and work to understand the requirements and expectations of the stakeholders who will be affected by their projects. They create a high-level design of the process flows that will be the focus of their project and determine the specific performance measures that will demonstrate the success of their project. Finally, the teams identify some quick wins that will help them to build enthusiasm for their projects and demonstrate to their champions and sponsors that they are committed and capable of achieving the outcomes specified by the sponsors. While the team members are doing their work, the champions are coaching and monitoring to ensure that the projects are being shaped in a way that fully aligns with the sponsors' intentions and that the teams are committed to using the DMAIC tool set. At this point, the champions have completed a successful handshake with their team members, and the green project teams are ready to begin their project work.

DMAIC, Blitz Teams, Organization-Wide Adoption

The teams work independently of one another, completing the activities specified in the DMAIC model. They are mapping the effected work processes, collecting data to determine root causes of the problem, analyzing the data to focus a solution on the root cause, and ultimately delivering a solution supported by a strong business case and proven through an effective pilot program.

During this phase, the teams are reporting their progress to the champions on a weekly basis. Champions are providing ongoing coaching and encouragement and communicating project progress to the sponsors on a monthly basis. Sponsors are providing feedback and actively supporting the teams by providing necessary resources and communicating progress and success across the organization.

The active, engaged sponsorship of the senior leaders and an active communications campaign prepare the organization to embrace the recommended solutions. Site-level blitz teams willingly sign up to implement the solutions at their sites, and organization-wide adoption begins.

Putting It All Together

Our goal in detailing how the integrated Six Sigma framework could be used to drive an organization-wide sustainability initiative was to illustrate how to integrate the use of high-performance teams, the Six Sigma methodology, and solid principles of change management successfully to achieve lasting change in a sustainability initiative. An inspired and engaged leadership team can leverage teams to get more work done faster. Through an integrated campaign, key messages are cascaded throughout the organization—from sponsors to champions, from champions to teams, and finally, from team members to peers across the organization. The principles of good change management are woven throughout the campaign. The result is measurable impact on the organization's sustainability objectives.

The Six Sigma Methodology Applied to Sustainability Projects

Now that we understand the Six Sigma integrated framework for team success, we can apply that framework to a typical sustainability team project focused on reducing energy consumption across a portfolio of data centers.

Background

Jones Lang LaSalle (JLL) has partnered with a global financial services institution to be its full-service corporate real estate services provider since 2005. Included in the portfolio of properties were six data centers with a geographic spread throughout the eastern United States, where a reduction in energy consumption of the site support infrastructure was targeted.

In February 2009, we formed a team composed of JLL energy, sustainability, and critical environment experts; the service leads from each of the JLL on-site engineering teams; the client facilities managers; and IT personnel. Team members met in a workshop format to lay out the project strategy, approach, and goals necessary to deliver a robust solution to reduce energy consumption without increasing risk to data center operations.

Define

During the initial workshop, the project was chartered, and the project goals were defined and communicated as follows:

- Consume less energy with no increased risk to the raised floor environment.
- Work quickly to deliver second half of 2009 impacts.
- Reduce overall utility usage in critical environments by 5 percent with a stretch of 10 percent.
- Achieve a $1 million reduction in 2009–2010 utility spend.
- Create a sustainable program for maintaining the energy gains.
- Create a methodology to ensure that the most effective and efficient practices are followed.

The project scope was focused on the energy needed to operate the data center infrastructure. While observations and suggestions could be made regarding the IT kit used within the data center, the primary focus of the project was the supporting infrastructure, including such items as utilities, heating/cooling, and lighting.

A Six Sigma master black belt managed the project, and a cadence of calls with the project team was scheduled to ensure consistent progress toward the identified goals.

Measure

One of the first Six Sigma tools used was the fishbone diagram. We used this tool for brainstorming factors that affected the energy demand and cost at the data centers. Through use of this tool, we identified the equipment and systems that are the major consumers of energy, the various set points that dictate the rate of consumption, and potential opportunities in the energy-rate structures and in the physical configuration of the sites.

The team then reached out to subject-matter experts for suggestions on successful energy-saving strategies for critical environments. We received a list of strategies, which the team discussed and pared down to those that fit into our project scope.

The team then set about to gather detailed information about each site to identify specific opportunities.

- A mechanical and equipment audit was conducted to understand the capabilities of existing equipment.
- A raised-floor audit was conducted to identify and correct air-management issues at each site.
- An analysis of utility rate structures was conducted to identify site-specific rate-setting trigger points and current performance.

Using utility bills, the team gathered data and compiled descriptive statistics and graphics to baseline energy-consumption performance by site (consumption and spend).

The define and measure activities up to this point in the project were focused on understanding the major drivers of consumption and rates, as well as identifying opportunities to shift the performance in a positive way.

Analyze

The analyze phase of the project was to focus and select the best of the identified opportunities so that the team could focus its efforts on the opportunities that would make the biggest difference. The cause-and-effect matrix tool was used to score each of the identified opportunities on impact to risk, consumption, energy rates, ease of implementation, and return on investment. As a result, the team focused on optimizing air-management techniques and temperature settings as the primary improvement opportunities; establishing rigorous modification testing methodologies to minimize risk.

Using the prioritized opportunities list, each site developed an action plan based on the specific situation at the site. Raised-floor tile configuration was optimized at each site to minimize total cooling demand and divert airflow to the areas of the data center generating the most heat. This action reduced the total amount of cool air required, enabling several cooling units to be shut down. As a control, temperature gauges were installed on individual racks and monitored to maintain a low-risk operating environment.

To further improve air management, a leak survey was conducted at each site, and leaks between the hot/cold aisles were eliminated, ensuring that the cooling was controlled and applied to the appropriate areas of the raised-floor environment.

With airflow controlled, the team focused on establishing the right set points for cooling in each of the sites. Temperatures were adjusted in half-degree increments and left to stabilize for several hours, with the effects being measured in the rack-mounted temperature gauges and metering of the site. When the optimal set point was reached that maximized the operating temperature and kept the fans on the individual servers from engaging, the consumption from air and temperature management was minimized.

Improve

Several quick-win opportunities presented themselves during the initial study, including the configuration and use of lighting at each site. More efficient lighting systems were put into place where applicable, and motion sensors were implemented so that the lights were on only when they needed to be.

Another quick-win opportunity was the recommendation to the IT organization that all new servers be ENERGY STAR–certified and that the energy-savings capabilities of the individual servers be turned on. While this was not adopted globally across the data-center portfolio owing to concerns about response speed with some critical applications, it did raise awareness of capabilities, which increased the use of energy-saving settings and contributed to revision of new server specifications.

The final quick-win took advantage of the configuration of one of the newer sites. When built, it was equipped with a 1 million gallon underground water tank. Given that the energy rates are significantly lower in the overnight hours, this tank was brought into service so that water, used to cool the site, could be chilled during the night at the lower rates. The water then would be used throughout the day, without additional chilling required, to keep the site at the specified temperature. The enabling of this capability shifted the entire cooling-energy demand to the cheaper overnight hours, saving over $150,000 per year.

Control

Estimated savings were calculated for each improvement opportunity that was implemented, but the predicted savings were validated at a site level using the trending and actual billing data to confirm the performance improvements. During the first year of the project, we reduced consumption at the six data centers by 4.9 percent, or $600,000, per year without increasing operational risk.

When we entered the second year of the project, we revisited and updated the brainstorming exercise and recalibrated the cause-and-effects matrix to highlight the biggest opportunities in 2010. We found that our biggest opportunity was to "operationalize" the changes made during the prior year by using the raised-floor tile, performing leak audits regularly, and updating the preventive maintenance and daily routines to monitor the settings and configurations so that they would adjust as the heat load changed.

We also identified some site-specific improvements that could be made to fine-tune the facilities and extract additional consumption savings. These items included changing out motors to higher-efficiency models, implementing better monitoring equipment, and using some newer technologies to reduce water consumption.

The second year of the project added an additional reduction of energy consumption by 3.6 percent. Total annual savings resulting from this project exceed $1 million per year.

Chapter Summary—Key Points

- Sustainability initiatives fail to achieve their desired results most commonly by failing to execute appropriately an effective change-management strategy and by failing to leverage the use of teams as a part of that overall change-management strategy.
- Despite conventional wisdom, even teams with a good understanding of the business situation, a reasonable level of authority, resources with which to take action, and an environment where they can think creatively often fail owing to deficiencies such as:
  - Uncertainty of purpose and lack of goal clarity
  - Narrow focus
  - Lack of authority
  - Insufficient data
  - Weak leadership
- The potential failure modes for team projects can be mitigated by the proper use of detailed team charters that are critically evaluated by business leaders, a supporting leadership structure, and the adoption of a consistent team problem-solving model known as DMAIC (define, measure, analyze, improve, control).
  - *Define:* Describe the project, business case, problem statement, goal, customer expectations, scope, and resources.
  - *Measure:* Detail the practical nature of the problem being addressed, define the project *Y*, define the data-collection plan, validate the measurement system, collect data for the *Y*, understand the process capability, and set quantified improvement targets.
  - *Analyze:* Engineer the problem as a statistical one, verify performance drivers (*x*'s), and identify performance gaps.
  - *Improve:* Shift focus to statistical solutions, select and pilot solutions, confirm improvements, map new processes, determine new capabilities, and develop an implementation and control plan.
  - *Control:* Confirm that the solution is sustainable, validate measurements, transfer ownership, and move to full-scale implementation.
- Effective solutions for improving enterprise sustainability require both solid technical design (e.g., through the DMAIC process) and solid change management. Ideas alone do not drive change. People drive change.

- In addition to describing each of the DMAIC phases, we elaborate on examples of deploying change-management tools such as GE's Change Acceleration Process and the Bridges model for managing transitions.
- The Six Sigma integrated team framework for driving transformational change includes the sponsors' launch, the champions' launch, the project launch, and widespread use of DMAIC and of blitz teams for fast-focus improvements.

Note

1. William Bridges, *Managing Transitions: Making the Most of Change* (New York: Perseus, 2003).

CHAPTER 6

Sustainability and Real Estate

As discussed in Chapter 3, real estate can have a significant impact on the environment in many ways, both directly and indirectly. Direct emissions are generated through material consumption, waste generation, water consumption, transportation, and the burning of fossil fuels for heat. However, the larger impact is through indirect emissions that are associated with the electrical power that is consumed by the building, with the emissions being generated far away at the power plant. According to the definition of the *Greenhouse Gas (GHG) Protocol*,[1] these emissions are defined as scope 2 emissions, as shown in Figure 6-1.

According to the Pew Center on Global Climate Change,[2] energy usage in buildings contributes 43 percent to national GHG emissions annually and up to 70 percent in some cities such as New York. Significant GHG emission reductions can be achieved from the built environment through the design and construction of new green buildings as well as the sustainable operation and renovation of existing buildings.

Additionally, buildings account for one-sixth of the world's freshwater withdrawals, one-quarter of its wood harvest, and two-fifths of its material and energy flows.[3] Building green is an opportunity to use resources efficiently while creating buildings that are protective of human health. Also known as *high-performance buildings*, green buildings result in increased productivity for building occupants, reduced operating costs, and minimized impacts to the environment. They reduce GHG emissions indirectly through energy (electricity and natural gas) efficiency and directly through resource conservation and recycling. When sited properly near public transit, they also can reduce transportation-related GHG emissions.

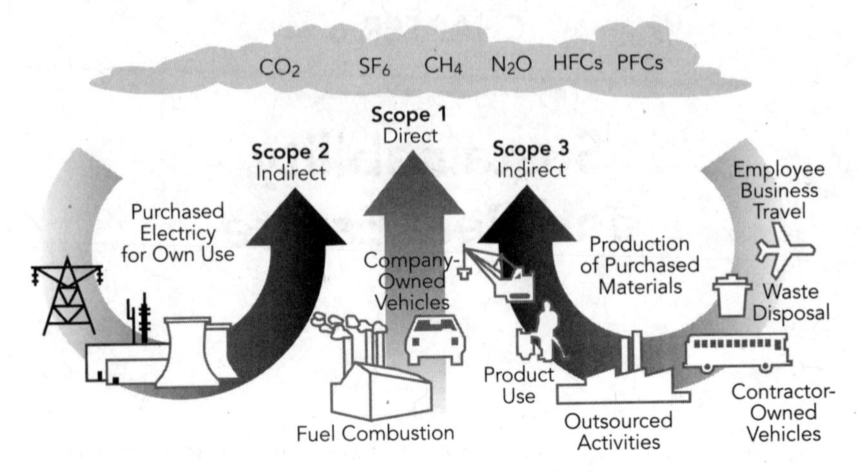

Figure 6-1 GHG emission definitions.

Moreover, while considerable opportunity exists for capturing GHG reductions during the construction of new buildings, the sheer size, comparatively low energy, and environmental performance of existing buildings suggests that improving these buildings must be a priority. It is estimated that in major American cities, 85 to 95 percent of all buildings that will be here in 2035 exist today, so if carbon-reduction targets are going to be achieved, it will require a focus on improving existing assets. The carbon footprint of companies in a service industry (not manufacturing and not transportation) is associated almost exclusively with the electricity and natural gas consumed by their buildings and with employee transportation—both commuting to work and business travel. Thus, for these companies, which in many cases have established aggressive goals for GHG emission reductions, reducing carbon associated with real estate is their only opportunity to have a meaningful impact. As shown in Figure 6-2, many companies are reporting their real estate carbon footprint according to the *GHG Protocol* categories labeled scope 1, scope 2, and scope 3.

While this reporting is focused on existing operating real estate, in reality, there are many decisions ahead of existing building operating decisions or new building design decisions that also can have a large and lasting impact on the environment.

Indicators	FY06	FY07	FY08	FY09
GHG EMISSIONS				
Total gross* GHG emissions Scope 1 (metric tonne CO_2)	27,586***	52,498	52,084	53,218
Total gross* GHG emissions Scope 2 (metric tonne CO_2)	317,686***	467,478	550,312	579,183
Total contractual* GHG emissions Scope 2 (metric tonne CO_2)	316,893***	403,188	310,961	226,733
Total air travel GHG emissions Scope 3 (metric tonne CO_2)	190,940	205,797	197,872	115,995
Change in air travel GHG emissions from FY06 (CGI global goal: 10% absolute reduction against FY06 baseline)		+8%	+4%	–39% (goal met)
Total contractual* GHG emissions Scope 1, 2, and 3 metric tonne CO_2)	535,419	661,483	580,917	395,944
Change in scope 1,2, and 3 from FY07 (EFA global goal: 25% absolute reduction against GY07 baseline*)			–15%	–40% (goal year is 2012)

Figure 6-2 Exampe of GHG reporting by scope.

Understanding that the overall goal (or "big *Y*") for corporate real estate is to provide a safe, comfortable, and productive work environment for employees at the lowest possible cost and with the least impact on the environment, this chapter focuses on the real estate decision criteria that will have the biggest impact on the environment. In its simplest form, this means adopting real estate decision-making criteria and practices that measure and reduce current and future emissions by reducing the square footage of space occupied and the emissions per square foot.

We also will touch on other real estate environmental considerations regarding the potential need to adapt to ongoing changes in climate and to consider and prepare for the actual physical risks of climate change to real estate assets and business operations.

Real Estate Decisions Relating to Energy and the Environment

Achieving meaningful reduction in real estate–related carbon emissions goes well beyond just building new Leadership in Energy and Environmental Design (LEED)–certified buildings[4] or improving our existing buildings. We need to reduce carbon dioxide (CO_2) per full-time equivalent (FTE) employee by factoring carbon, energy, and climate risk into every real estate decision. A LEED Platinum building is not the answer if it accommodates one person per 300 square foot or if people have to travel a long way to get there. The greenest building is the one that you don't build because of improved occupancy strategies and workplace standards. A green building or green real estate program needs to consider how design, operational, location, and investment decisions will affect a broad range of factors and needs to quantify those dimensions so that one building can be compared with the next, as indicated in the spider diagram in Figure 6-3.

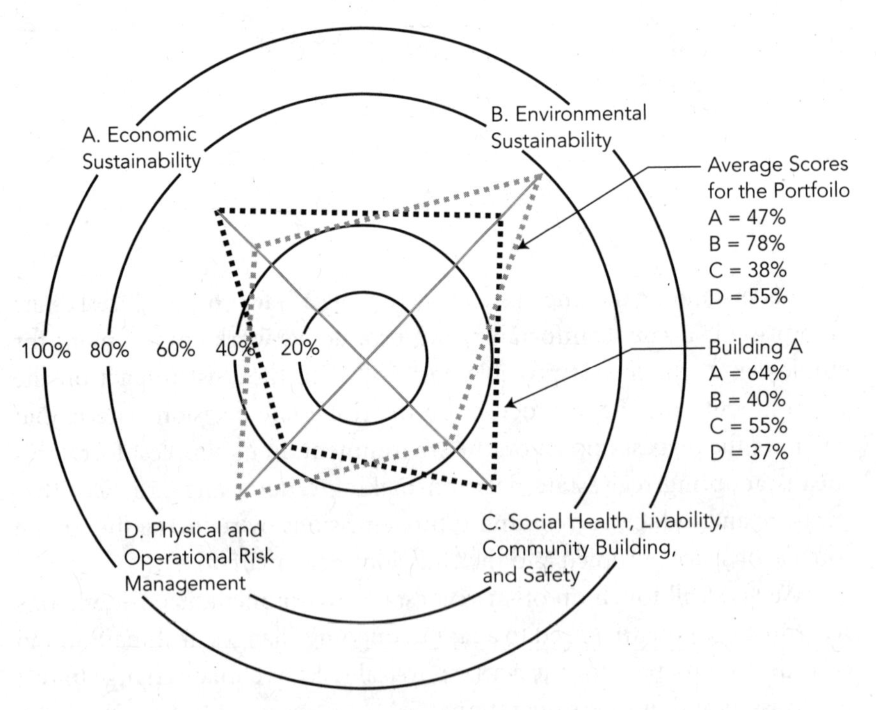

Figure 6-3 Spider diagram.

And while the primary focus may be on reducing carbon emissions and energy consumption, green building strategies and alternative work environments are proving to generate many other benefits to corporations:

- Met corporate goals for emission reductions
- Lowered utility expenses
- Increased employee attraction/retention and productivity
- Improved regulatory compliance
- Increased customer loyalty
- Improved shareholder satisfaction
- Reduced physical risk
- Provided more flexible and collaborative work environments

The Right Steps in the Right Order

To understand how to best sequence the actions required for a more sustainable real estate portfolio, we turned to Jones Lang LaSalle. Jones Lang LaSalle is a global leader in real estate services and financial management. The company owns or manages one of the world's largest portfolios of commercial property and has been recognized as a leader in sustainability. According to Jones Lang LaSalle, there is a natural progression associated with minimizing the overall environmental impact of corporate real estate in terms of potential impact beginning with the concept that the greenest building is the one that you don't build because you are using other existing real estate more efficiently:

1. Companies should seek to reduce their overall real estate needs in terms of total square footage.
2. Location decisions should be made that will minimize the environmental footprint for the real estate that is required.
3. Necessary new buildings or space in existing buildings should be designed and built to meet or exceed current green building or high-performance building standards.
4. Existing buildings or space should be upgraded, where appropriate, and then operated as efficiently as possible.
5. Employee engagement programs should be implemented to enlist ongoing employee support in behaviors that will contribute to a reduced environmental footprint.

However, while these steps may be the ideal order in terms of total impact, Figure 6-4 shows that initiatives that will have the largest impact often require the longest time to implement and thus may lead to an inverse relationship in terms of timing of the implementation. In reality, any comprehensive program should consider attacking multiple categories of opportunities in parallel. This approach does require close and careful consideration of a myriad of ongoing factors relating to changes in occupancy needs driven by changes in business requirements, other capital requirements for existing buildings, and the availability of financing to support certain strategies. An example of how the lack of a comprehensive approach can lead to a suboptimal answer is described in the Apex example that follows.

Returning to our Apex case study, Apex had multiple properties in a central business district and was very focused on identifying and pursuing energy-conservation improvement opportunities in its buildings. The company launched an effort to perform energy audits to identify both

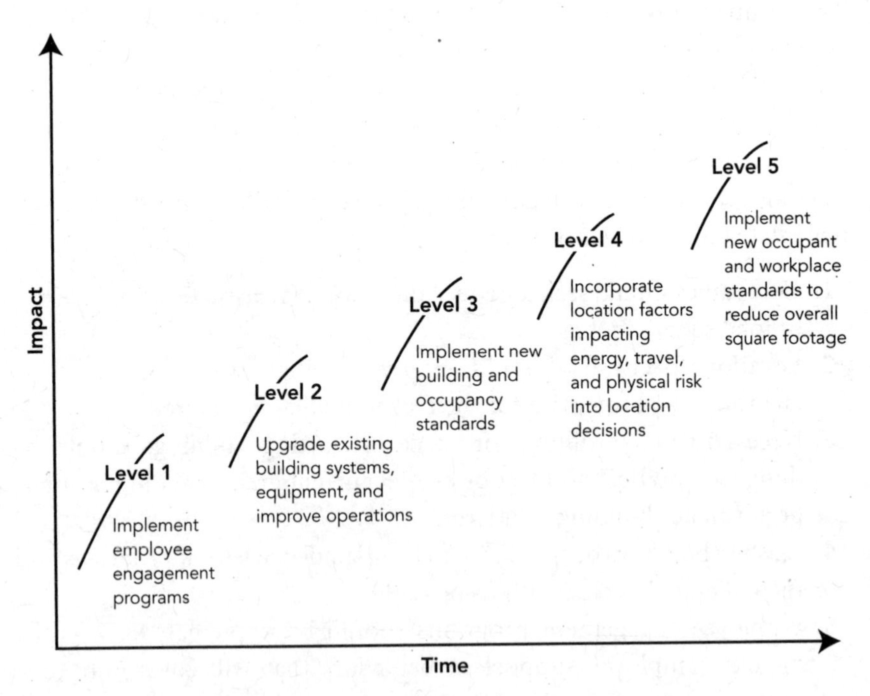

Figure 6-4 Corporate real estate actions.

operational improvements and potential capital upgrades that would deliver a reasonable financial return. Two buildings in particular appeared to be good candidates for both operational changes and capital upgrades to the lighting and heating, ventilation, and air-conditioning (HVAC) equipment, among other miscellaneous improvements. One building was a high-rise office building, and the other was a nearby, smaller office location that primarily housed some back-office operations. The high-rise office building also had a lot of vacancy scattered throughout the building on various floors. The smaller building had a number of very attractive potential energy-conservation projects that would require some upfront capital but would deliver a fairly short payback. Taken in isolation, it was likely that the engineering team would gain approval to move ahead with the capital upgrades in the smaller building. However, when the local market scenario was evaluated from a more holistic real estate perspective, the corporate real estate team soon realized that by implementing its revised workplace standards in the larger building, it could restack the floors to increase density and consolidate the vacant space in a way that they could accommodate more than 100 percent of the workers from the smaller building. At the same time, the team could implement some of the energy-conservation measures in the larger building and could completely dispose of the smaller building. Using this more comprehensive review enabled the Apex team to realize a reduction in total square footage and to prevent unnecessary expenditures at the smaller building.

Reduce Your Space Requirements

It begins with the simple premise that the easiest and most effective way to reduce your real estate–related environmental footprint is to reduce your overall real estate footprint or square footage. In other words, the greenest building is the building that you do not build because you are using space efficiently or are finding alternative work locations for your people. Alternative sites for employees may include working from home, working at a client location, or sharing offices. Using space efficiently generally involves revising your office standards to increase the density of the workstations to accommodate more people in less space. The common approach is to eliminate private offices in favor of a more open work plan. This design approach has the added benefits of improving employee collaboration and improving the penetration of natural daylight into the office areas, thus reducing the need for electrical consumption while providing an improved office environment. Companies with many employees who travel frequently or spend significant time at client locations have had success in increasing density through desk-sharing programs. Some programs have gone so far as to have ratios of 15 employees for every workstation, although a ratio of four to one is much more common. These ratios depend greatly on the type of work or the industry—with sales, information technology (IT), and other mobile workforces having the higher ratios.

Another common measure of space efficiency is total square footage per employee, with 50 to 100 square feet per person becoming standard for leading-edge companies. Simple math shows that a company with 10,000 employees at a current standard of 350 square feet per person (which is not uncommon) would need 3.5 million square feet of office space. Migration to standards that would provide 150 square feet per person (not even best in class) would generate only 1.5 million square feet in space needs, or a reduction of 2 million square feet. With an average carbon consumption of 22 pounds per year per square foot, this improvement in space efficiency would translate into a reduction of approximately 20,000 metric tons of carbon annually, which is the equivalent of 3,824 passenger vehicles per year or the annual energy consumption of 1,702 homes.

There are numerous references and studies that provide much greater detail on the methods and processes for improving office density with consideration for balancing privacy and collaboration and for managing some of the human resources, change-management, and employee communications implications associated with employees working from other locations such as home.

Factor Carbon into Location Decisions

Real estate location decisions can have significant and long-lasting impacts on a company's total environmental footprint in many categories, including transportation, energy and water consumption, electrical power sources, site selection, and physical risk of climate change.

Transportation

The physical location of a company's offices can affect employee commute miles, business travel, and shipping and receiving–related transportation. For instance, locating offices close to public transportation or close to where the employee base lives can minimize employee commutes, locating offices close to clients or close to major transportation hubs can have an impact on business commuting, and locating offices near ports, rail centers, and so on can have an impact on shipping and receiving considerations, particularly for manufacturers or distributors.

Energy and Water Consumption

Location decisions also should consider the potential impacts of climate on energy and water consumption. Most office environments require more cooling than heating, so as global average temperatures continue to increase, some companies are beginning to favor more northern climates (e.g., in the northern hemisphere), which can reduce the environmental cooling load on the building. These considerations also must be weighed against other location factors, such as availability of alternative or low-carbon energy sources. For instance, solar-energy opportunities in southern climates may outweigh the reduced-load considerations of northern climates.

Electrical Power Sources

A third location consideration, also described in Chapter 3, relates to the carbon associated with local power production. A company's carbon footprint can be affected by the power source of the electricity the company buys. Power that is generated by coal-fired plants has a higher carbon factor than power generated by hydroelectric or nuclear sources. Thus, locating your facilities in largely coal-fired utility districts will increase your carbon footprint, with all other things being equal.

Site Selection for New Space

Another location consideration for new construction relates to brownfield versus greenfield sites. A *brownfield* site generally is defined as reuse of an existing site, often in an urban environment, which may involve refurbishment of an existing structure or rebuilding on an existing site. A *greenfield* site involves new construction on a site that has never been developed. From a lifecycle perspective, it is always better to locate your offices in an existing facility, but if you must consider new construction, urban infill locations that take advantage of existing water, sewer, power, and street infrastructure are highly preferable to greenfield locations in rural or suburban areas that require new infrastructure and may disrupt plant and animal habitats.

Physical Risk of Climate Change

A final and related location criterion for many firms, especially when considering placement of mission-critical facilities, is the physical risk of climate change. Insurance companies are also paying close attention to this topic. The factors to consider here include the increased threat of extreme weather events such as hurricanes and floods and the potential for rising sea levels and the impact of these events on power availability or reliability and transportation. While these elements will not necessarily affect your carbon footprint, they are certainly factors that many firms are beginning to consider in their location decisions. Many firms, for instance, would not locate a mission-critical site such as a data center in the Gulf Coast region of the United States. With climate change threatening to increase the frequency and severity of humanitarian crises, economic disruptions, and weather-related disasters (which in the last half-century have cost more than a trillion dollars and have killed more than 800,000 people), insurance companies and others are paying close attention to this issue. The Intergovernmental Panel on Climate Change (IPCC) recommends that businesses should consider the physical effects of climate change, which may include a wide range of geographic, market, and sociopolitical factors. As a starting point, the IPCC suggests that businesses focus on the most obvious and intense business impacts likely to result from extreme weather, especially in coastal and floodplain regions. Many of the lasting business impacts of recent flooding events in the United States have been related to disruptions to transportation because of damage to highways, bridges, and rail lines.

Some companies are taking obvious actions. Following Hurricane Katrina, Entergy, a power company located in the Gulf Coast region, took immediate action to relocate important business centers, including moving a data center to Little Rock, Arkansas (creating redundancy in data storage throughout the service area), and moving its transmission center to Jackson, Mississippi. Entergy made decisions about where to locate these important business centers based, in part, on information about the climate-related risks of the different geographic regions within its service area and in order to locate centers and buildings in different parts of its service area. In addition, Entergy put together a business continuity group specifically tasked with looking at broader implications of climate in the context of other serious business threats, including terrorist acts and a potential flu pandemic.

Apply Green Standards to New Buildings or Space

When a new building emerges as the right choice for new office requirements, and after location decisions have been made mindful of the considerations outlined previously, then it is key to incorporate as many of the green building or high-performance building standards into the process as is possible and practical. These standards, such as the Leadership in Energy and Environmental Design (LEED) standards from the U.S. Green Building Council, are becoming very mainstream for new construction in the United States and even around the globe. Early concerns about first-cost premiums associated with building to these standards have virtually disappeared as architects and engineers have learned to apply them and as contractors have learned to incorporate construction techniques that consider local sourcing, construction debris management, site-conditions management, and so on. Also, the cost and availability of recycled and recyclable materials have improved as market demands have increased. At this stage, it would be naive to not consider these design guidelines, even if LEED certification is not a goal.

The benefits of incorporating these design concepts have been well documented and are numerous, including

- Reduced utility costs for electricity, water, sewer, and natural gas
- The associated reduction in carbon footprint
- A healthier work environment with respect to air quality
- A more comfortable work environment with respect to lighting and thermal comfort

Studies, such as the recent Michigan State study on the effects of green buildings on employees, have shown that work environments that incorporate green building design concepts contribute to reductions in employee turnover and absenteeism.[5] Green building designs incorporate a wide range of design concepts and approaches that minimize the overall environmental impact of a new building and go well beyond just carbon footprint and employee comfort and productivity. Common design evaluation components include

- Site conditions:
  - Site selection—greenfield versus brownfield
  - Proximity to public transportation
  - Building orientation on the site to maximize natural daylighting and minimize solar heat gain
  - Construction techniques to minimize site erosion, storm water runoff, and so on
- Water:
  - Selection of fixtures to minimize consumption
  - Use of rainwater collection or graywater or blackwater systems
  - Landscaping that minimizes the need for irrigation
- Energy:
  - Building envelope design and materials to minimize infiltration, solar heat gain, and thermal loss while maximizing natural daylight
  - Use of alternative energy sources (e.g., wind, solar, or geothermal)
  - Use of high-efficiency HVAC and lighting systems
  - Advanced lighting and HVAC control systems that respond to building demands and availability of daylight and allow for remote monitoring and demand control to take advantage of utility company incentive programs or future access to the smart grid
- Materials:
  - Reuse of existing materials and use of recyclable materials
  - Minimization of construction debris
  - Sourcing of materials locally
- Indoor air quality: HVAC designs to use and deliver the right amounts of outside, filtered air to occupied spaces based on actual demand

When it comes to the build-out of new leased space in an existing building, many of the same design concepts should be incorporated, but there are other considerations to be addressed ahead of the actual design phase. The first step, as discussed previously, is to select a location that minimizes impacts relating to transportation, climate change, and so on. Once a market or submarket is selected, then an additional challenge is to select one or more candidate buildings that best support your new green office standards. Design elements that should be incorporated into your evaluation of candidate buildings include

- *Base building design.* Is the building built to LEED or other green building standards?
- *Base building performance.* Is the building being operated to maximize the benefits of the original design? (It is not uncommon for buildings designed to high-performance building standards to be operating far below their potential because of issues with respect to the original building testing and commissioning or with respect to poor ongoing maintenance and operating practices.)
- *Building operating practices.* Is the building manager incorporating green building operating practices that go beyond the building infrastructure and design and include programs such as green cleaning, green landscaping practices, pest control, recycling programs, and so on?
- *Support of space standards.* Will the physical floor plate allow you to incorporate your office and furniture standards that have been chosen to maximize density? (Some base building floor plates and curtain-wall designs lend themselves better to open office designs than others.)
- *Natural daylighting.* Will the existing window and curtain-wall system enable you to reduce interior lighting in response to the availability of natural daylight? (Clearly, your ability to maximize the use of natural daylight in an existing building is determined largely by the base building design.)

On identifying one or more buildings in the right market to meet your green building standards, there also is a need to incorporate certain ongoing elements and operating practices into the lease. A green lease can be used to protect you as a tenant by placing certain obligations on the landlord to maintain and operate the building to the design standards and to use green operating practices. A more comprehensive discussion on green leasing programs is included later in this chapter.

Existing Building Operations and Retrofits

The fourth category to address when improving the environmental performance of your real estate is your ongoing building operating practices, which include the operations of whole buildings that you own and occupy or your operations in leased space. For new buildings that have been designed to green or high-performance building standards or for existing buildings that have been upgraded to these standards, the key to success here is to be sure that the building continues to operate according to the original design intent and/or that appropriate modifications are made in response to changes in occupancy or use. If the original design was correct, maintaining the building to the design standard is a matter of ensuring that systems and equipment are operating only when absolutely necessary and then, when they are, that they are operating efficiently. This usually requires continuous monitoring of the building systems and equipment along with good maintenance practices to be sure that equipment components (i.e., fans, chillers, cooling towers, dampers, valves, pumps, motors, and so on) are operating properly. On office floors or in leased space where you may not control the base building systems or equipment, good operating practices again start with making sure that lighting, office equipment, and personal computers are powered on only when absolutely necessary. Ideally, there are automated controls in place to handle these simple on/off functions, such as occupancy sensors in conference rooms (or other special-purpose spaces) and personal computer power-management systems.

For existing buildings, changes in occupancy or use or poor maintenance practices often have led to a dramatic deviation from the original operating design, causing the building systems and equipment to operate very inefficiently. Temperature set points may have changed, or control devices such as thermostats, valves, and dampers may be out of calibration. These buildings often are great candidates for a practice known as *retrocommissioning*, whereby the entire control system and the associated control devices are recalibrated. Many other existing buildings are great candidates for a more extensive retrofit or refurbishment. Advances in technology and increases in energy costs create opportunities to upgrade or replace building systems and equipment with an attractive return on investment. Common energy upgrade initiatives include

- Replacement of motors on pumps and fans with new high-efficiency motors
- Replacement or upgrade of chillers, boilers, or packaged HVAC systems
- Replacement of existing lamps and ballasts in light fixtures or complete replacement with new fixtures
- Converting existing incandescent lighting to compact fluorescents or light-emitting diode (LED) lamps
- Implementing lighting controls to sense and adjust for occupancy and natural daylight
- Adding insulation
- Upgrading the control and automation system
- Adding alternative-energy supplies such as solar or wind

A notable example is the Empire State Building, where the implementation of some relatively standard energy-conservation initiatives will enable an older 1930s building to become a modern, high-performance building. The upgrades include retrofitting the chillers, adding new building automation, refurbishing the existing double-pane windows to add a reflective film and argon gas, adding lighting controls, upgrading the fan systems, and adding tenant power-measurement and control capabilities.

Operating the building or your space within a building to design standards primarily addresses energy consumption. From an operating perspective, other programs to consider include

- *Green cleaning.* Green cleaning programs are becoming much more common and incorporate new standards for the quantity and types of chemicals that are used, the selection of restroom paper products, and even the time of day for cleaning. Some firms have converted to daytime cleaning in office spaces that include the use of new high-efficiency vacuums that are very quiet and are not disruptive to office workers. An added benefit of this approach is that it also minimizes the need to light office spaces after hours for the cleaning crew.
- *Green landscaping practices.* The selection and maintenance of indoor and outdoor plants can be made to reduce the need for watering and for chemical pesticides and fertilizers.
- *Green pest control.* New pest-control programs include reducing or eliminating the use of harmful chemicals.

- *Reduce/reuse/recycle programs.* Office waste-management and recycling programs are an obvious opportunity to reduce your environmental impact. The right approach is first to reduce your consumption, then to reuse materials whenever possible, and then to recycle. The use of office paper is an easy example. Encouraging employees not to print materials unless they absolutely must and then to print only double-sided can reduce consumption. A similar savings can be obtained when employees are encouraged to reuse any paper that has been printed on only one side for draft printing or scratch paper. And finally, all paper that is being discarded should be recycled. However, office paper is only one small part of the recycling opportunity, which also should include cans, bottles, ink toner cartridges, batteries, and electronics such as cell phones, personal computers, and monitors.
- *Water programs.* Common water-reduction programs include the replacement of existing faucets and toilets with low-flow fixtures or waterless urinals. Properties with landscaping replace formal, irrigated lawns and beds with native plantings that do not require irrigation.
- *Building mechanical maintenance practices.* Ensure that an ongoing preventive and predictive maintenance program is in place to maintain all building systems and equipment in peak operating condition and to avoid any future requirements for retrocommissioning. Program building automation systems or monitor real-time building and system performance to optimize performance in response to changes in weather and occupancy.

Engage Employees in Support of Green Initiatives

The topic of employee engagement is covered extensively in Chapter 5, so this section focuses on elements of an overall employee engagement program that have a direct impact on the environmental performance of offices and real estate. The prior discussion on building operating practices focused primarily on operations of the base building infrastructure. However, how employees use the space, systems, and equipment in the building also can have a dramatic impact on total energy consumption, material consumption, and recycling programs. Employee engagement represents a real opportunity to accelerate progress toward sustainability goals, but it is often overlooked. Unfortunately, many companies are still missing this opportunity to engage fully their most powerful resource—their employees. Every day, employees across an organization make decisions with far-reaching environmental consequences. Informed decisions about such straightforward matters as programming heating and cooling systems, adjusting energy settings on computers, commuting to work, and even purchasing paper, printers, and kitchen equipment can reduce a company's carbon footprint and save money. When it comes to fast ways to make a big dent in both your greenhouse gas emissions and your energy bill, personal computer (PC) power management is one of the easiest actions a company can take. Most power-management software firms predict $20 to $60 in savings per computer per year from reduced energy use. For large companies, these savings can add up very quickly. AT&T, for example, launched a PC power-management program that will save the company more than $13 million a year.

At their best, employee engagement programs can help to deliver on low- and no-cost programs that stimulate employee participation and potentially save millions of dollars annually in energy and other resources, not to mention helping to reduce a company's carbon footprint dramatically. Motivating workers toward environmentally friendly behavior also rewards companies in other ways. For instance, studies show that employees in sustainability-focused workplaces are 6 to 16 percent more productive, and research has shown that 90 percent of people now say that they wish to work for a company with a strong green reputation. In a recent study by Johnson Controls,[6] 96 percent of Generation Y employees said that they want an "environmentally aware or friendly workplace."

Many times the key to success is to provide enough education to enable employees to make informed decisions and then let their own motivations take over. Many of the best ideas come from people at the worksite, such as a Walmart employee who suggested turning off the lights in break-room vending machines, a move that now saves nearly $1 million each year in electricity cost across the company. It is also useful to tie into well-known global events such as Earth Day and Earth Hour or engage with local environmentally focused groups. Procter & Gamble employees have the opportunity to complete an online examination of their green office habits, which not only scores their behavior but also generates a detailed personal action plan for improvement based on their answers. Jones Lang LaSalle conducts a mandatory "Sustainability in the Workplace" global e-learning course to ensure that all employees can articulate the key components of the firm's sustainability commitment, as well as demonstrate basic knowledge of green best practices within the real estate industry.

Green Leasing

As mentioned in the section on real estate location decisions, identifying a green building to occupy and building out space to new standards are not sufficient if the landlord can allow the building to deviate operationally from the design standards or is allowed to employ poor environmental operating practices. Corporations that lease the space they occupy rely on the participation of building owners to help them achieve their sustainability goals. This requires some types of "green lease" clauses.

Contrary to popular perceptions, there is no one-size-fits-all green lease document. In the most basic sense, a green lease is any lease to which some agreed-on sustainability concepts have been added. However, for a green lease to be effective, it must

1. Assimilate the tenant's specific sustainability goals for the space
2. Outline mechanisms for measuring performance to ensure that goals are achieved initially and throughout the lease term
3. Provide for appropriate allocation of costs for achieving agreed-on sustainability standards
4. Offer the tenant appropriate remedies in the event the landlord fails to perform his or her sustainability obligations

Some key principles that should be part of the overall lease discussion include

1. The landlord should operate the building and the tenant should operate its premises as efficiently as feasible.
2. For any given system, installation, or piece of equipment, responsibility for capital expense and benefit of savings should reside with the same entity.
3. To the extent feasible, both consumption and demand for resources throughout the building should be measurable and transparent to both landlord and tenant.

We strongly recommend that tenants develop specific sustainability standards before developing a green lease. These standards will guide the negotiations and ensure that the lease stipulates both the standards themselves and how they are to be implemented.

Further detail on what to look for in a green lease includes

1. Does the lease specify general building standards? To achieve its sustainability goals for its leased space, a tenant often will want the building it occupies to meet specific standards (e.g., HVAC performance, availability of renewable energy, a building-wide recycling program, use of environmentally friendly janitorial supplies and practices, and so on).
2. Does the lease specify the standards to which the landlord will deliver the premises? Obviously, when the landlord is to build out the premises, the plans and specifications must reflect the tenant's sustainability objectives. However, even when the tenant plans to manage the build-out, the raw or partially finished premises and the supporting building systems must be suitable to achieve the tenant's sustainability goals.
3. Does the lease require the landlord to maintain specified sustainability standards for the duration of the lease term? It is not sufficient for the building to meet a specified sustainability standard on the delivery or commencement date. The building should continue to meet or exceed the standards throughout the lease term. The typical lease language that states that the landlord will maintain and operate the building in a "first class" manner or in a manner achieved by other "comparable" buildings in the market is not sufficient.
4. Is the building LEED certified or has it received a green rating? LEED certification and other sustainable designations such as ENERGY STAR or Green Globes are desirable and often indicate that a building will meet a tenant's sustainability criteria. However, there are a number of ways to achieve certification, so the lease should specify in detail the standard to which the building will be built or upgraded. Furthermore, the certification process is complex and requires considerable lead time; therefore, the lease should establish milestones for achieving certification and penalties for failure to meet them (e.g., a rent reduction).
5. Are you planning to pursue LEED certification or another designation for the occupied space? If the landlord has no intention of pursuing LEED status, it is still possible to attain LEED or other certifications for

the occupied space. In such cases, the landlord must provide historical and current data on the building's energy use and operational practices in order to demonstrate compliance with certification requirements. Additionally, the tenant should be permitted to provide a copy of the lease for such purposes.

6. Does special monitoring equipment need to be installed? In order to measure system performance and ensure that building/premises standards are met on an ongoing basis, it may be necessary to install, maintain, and operate special monitoring and data-collection equipment. The obligation and cost allocation are negotiable, but the tenant must have access to the data either on a continuous basis or via periodic reports that cover the tenant's overall sustainability metric requirements.
7. Does the lease give the tenant self-help rights if the landlord fails to achieve/maintain the established sustainability standards? Many leases give the tenant limited self-help rights if the landlord fails to perform certain obligations (e.g., repairs and maintenance within the premises). Typically, such self-help rights do not extend to the landlord's obligations beyond the premises itself. In a green lease, it is imperative for these obligations to be addressed, including maintaining sustainable practices relating to the building systems and the building itself. Because the tenant specifically negotiated these services, self-help rights are necessary to ensure that the terms are met throughout the duration of the lease.
8. Does the lease permit the tenant to perform *fundamental commissioning* of the building systems? In order to achieve LEED certification, the tenant must have the building systems (e.g., HVAC) fundamentally commissioned to ensure that they are installed, calibrated, and perform as intended, and therefore, this must be addressed in the lease.

Some key things to avoid in a green lease include

1. An assumption that LEED or other certifications alone are sufficient (These designations do not necessarily mean that a building meets a tenant's sustainability requirements. LEED has four levels of certification: certified, silver, gold, and platinum. The level is determined by point totals based on criteria outlined by the U.S. Green Building Council. The party seeking certification can select which

requirements or points to pursue. These points may or may not reflect the objectives of the tenant, and therefore, the tenant should spell out the specific areas of focus required.)

2. An obligation to comply with a vaguely defined sustainability standard that is within the landlord's discretion (Landlords are becoming increasingly aware that green buildings offer a competitive advantage in terms of both attracting potential lessees—which may translate into higher net rents—and cost savings through energy reduction. Landlord-oriented lease forms (e.g., the *BOMA Commercial Lease Guide*[7]) are emerging now that call for the cooperation of tenants in sustainability practices, which are undefined and left to the landlord's discretion. Because cooperation presumably would be at the tenant's cost, you should insist on specificity of any such obligations.)
3. Any landlord attempts to capture carbon offset credits and other sustainability incentives relating to the building (Both governmental regulations and private markets are creating sustainability incentives such as carbon offset and tax credits, which can become an important financial component of any green building or lease. Landlords may attempt to capture these credits and incentives for themselves or seek discretion over their allocation. You will want to capture any credits and incentives that relate to your premises and make sure that the building-related credits and incentives are used to reduce operating expenses and/or taxes and passed through to building tenants.)
4. Landlords allocating building-wide sustainability costs on a non–pro rata share basis (If you are the first tenant in the building that is requiring building-wide sustainability practices in its lease, the landlord may try to impose the costs on you rather than passing them through to all tenants on a pro rata share basis. If it is an existing building, the landlord may assume that he does not have the right to pass such costs to existing tenants or, if it's a new building, that other tenants may not care about sustainability and won't want to pay the additional cost. However, the landlord will reap the benefits of having a green building, as will the other tenants, and therefore, all should share the cost burden as well as the benefits.)
5. A requirement that the tenant obtain its utilities from the landlord or through a landlord-designated supplier (With the advent of utility deregulation and the availability of renewable-energy providers, it is

critical that the tenant be able to select his own energy suppliers to reduce both costs and carbon footprint.)

6. Time limits on the landlord's obligations to meet the agreed-on green standards (Landlords may complain that they can only guarantee higher performance initially or during the new equipment warranty period owing to the natural degradation of equipment performance over time. This is not acceptable. The issue often can be resolved by permitting the landlord to include, in annual operating expenses, the amortized cost—over its useful life—to replace equipment that no longer meets the higher standards if normal maintenance and calibration will not suffice.)
7. Conflicting rules and regulations (It is common for the rules and regulations attached to a lease to address matters, including sustainability, that have been negotiated in the body of the lease. Be sure that any rules and regulations that are redundant or in conflict with the lease itself are deleted or conformed and that the mechanism to amend the rules and regulations does not permit the landlord to make unilateral changes to lease provisions in a manner adverse to the tenant.)

Other green lease clauses or considerations should include

1. *Hazardous materials.* This lease clause should define what is a hazardous material and that neither the landlord nor any tenant violates any federal, state, or local laws or regulations regarding the presence, disposal, storage, generation, or release of hazardous materials.
2. *Green cleaning specifications.* This lease exhibit should define the materials, procedures, and protocols for cleaning the building in a sustainable manner.
3. *Building rules and regulations.* This lease exhibit should contain language to stipulate a building-wide recycling program. Day-to-day operational issues can be addressed in a "Recycling Program Guide" that is cited in the rules and regulations.
4. *Tenant construction agreement.* This lease exhibit should contain language to define sustainable product requirements and construction practices to be used in constructing the tenant improvements.
5. *Tenant Manual and Development Guidelines.* This guide would explain the building's sustainable features and benefits, procedures, and

operating parameters in nontechnical terms to tenants. The development guidelines should provide the tenant with insights into how to maximize the building's design features and systems to create a high-performance and sustainable workplace.

Financing Multitenant Building Retrofits and Existing Lease Constraints

Pursuing capital upgrade projects in your owned facilities is relatively straightforward and often begins with some type of green building assessment or energy audit to identify opportunities to improve overall performance and to reduce energy consumption. Typical upgrade or retrofit opportunities include such things as lighting retrofits, HVAC upgrades, control-system enhancements, and building-envelope repairs and upgrades (e.g., doors and windows and alternative-energy installations). The energy reduction and potential financial payback for these opportunities are usually clear-cut as well and enable these projects to be evaluated relative to other capital program expenditures as part of a corporation's overall capital planning process. In an investor-owned multitenant building, the approach to capital upgrades can be much more complex. An investment that appears to be a short 2- to 3-year simple payback can quickly turn into an 8- to 10-year payback owing to the structure and constraints of existing leases. Let's say that a $1 million dollar central plant is calculated to deliver $250,000 in annual operating savings for a simple 4-year payback. Most leases will make the landlord responsible for the $1 million upfront cost, whereas all or a portion of the savings (depending on gross versus net versus modified gross leases, expense stops, and so on) will accrue to the tenants. The 4-year payback can quickly become a 10-year or longer payback even when accounting for tenant turnover and depreciation. This issue is known as the *split incentive* because it causes the landlord to be responsible for the cost of the capital upgrades, whereas the tenants receive most of the cost-saving benefits. Forgetting issues with respect to ensuring that the property continues to be cost competitive with others in the local market, if owner capital is available, it is likely to be invested more profitably elsewhere. A 2007 report by the Lawrence Berkeley National Laboratory[8] on the growth and development of the energy services industry specifically identifies the issue of the split incentive as a barrier to industry growth. Owners and tenants alike are beginning to recognize that new green leases present new opportunities to address this issue and better align landlord and tenant incentives.

Measuring Success and Environmental Reporting

Carbon reporting and measurement have become new challenges for companies as they attempt to measure progress against their environmental goals, issue environmental or corporate social responsibility (CSR) reports, respond to stakeholder questions, and consider other reporting or listing opportunities such as the Dow Jones Sustainability Index (DJSI), the FTSE4GOOD Index Series (FTSE4GOOD), or the Australian Sustainable Asset Management (SAM) Sustainability Index (AuSSI). The Global Reporting Initiative (GRI) provides a Sustainability Reporting Framework incorporating Sustainability Reporting Guidelines that are used by some 1,000 organizations worldwide. Additionally, the global Carbon Disclosure Project (CDP) specifically petitions organizations to publicly disclose their carbon-emission performance. Environmental performance-rating tools are becoming widely used and recognized in the industry, with an increasing focus on results at a property level. New regulations are emerging in countries, cities, and states that require different types of reporting or levels of disclosure. For instance, the European Union recently enacted requirements for building-related carbon reporting known as Energy Performance Certificates that require owners to publicly display the performance of the building. The State of California recently passed legislation requiring building owners to produce an ENERGY STAR score for any building that is being sold or master leased. Meanwhile, a number of building rating and evaluation schemes exist around the world, including LEED in many countries, Green Globes in the United States and Canada, BRE Environmental Assessment Method (BRREAM) in the United Kingdom, National Australian Built Environment Rating System (NABERS) and Green Star in Australia, and others. Organizations are now faced with the challenge of balancing these sustainability reporting requirements, which at times are complex and varied, with existing operational performance reporting and analysis. Environmental reporting, once it is codified in legislation, may well place the same duty of care on company directors and finance teams as financial reporting requirements do today. As such, fully integrated reporting and analysis systems are likely to be instrumental for corporations around the globe to manage and verify sustainability reporting. If you are not measuring and reporting your real estate–related carbon today, it is likely that regulations or customer/shareholder pressures will create the need for reporting in the future. Any new reporting program should follow the emerging global standard protocols and use a system and process that can be audited and are repeatable.

Chapter Summary—Key Points

- Buildings account for large portions of global GHG emissions, freshwater withdrawals, wood harvest, and material and energy flows. Building green is an opportunity to use resources efficiently while creating buildings that are protective of human health.
- Although considerable opportunity exists for reducing GHG emissions by constructing new, more efficient buildings, the sheer size of the existing building portfolio suggests that improving existing buildings must be a priority.
- Real estate decisions in support of environmental sustainability go beyond resource efficiency—building a LEED-certified building is not the answer if it accommodates one person per 300 square feet of office space or if employees must commute from far distances. The greenest building is the one that you don't build because improved occupancy strategies and workplace standards have you using your existing portfolio more effectively.
- There is a natural progression to minimize the environmental footprint of a real estate portfolio:
  - Reduce space requirements.
  - Factor carbon into location decisions.
  - Apply green standards to new buildings or space.
  - Address existing buildings and retrofits.
  - Engage employees in support of green initiatives.
- Corporations that lease the space they occupy must rely on the participation of building owners to help them achieve their sustainability goals. Green leases can be used to
  - Codify a mutual commitment to operating the building efficiently

- Align financial incentives for improving efficiency
- Improve transparency in environmental performance of the building

- Companies not measuring and reporting real estate–related carbon footprint today should anticipate that regulations and/or customer expectations for doing so are likely future scenarios.

Notes

1. Pankaj Bhatia and Janet Ranganathan, *The Greenhouse Gas Protocol: A Corporate Accounting and Reporting Standard*, rev. ed. World Resource Institute; Washington, DC (March 2004). Available at www.wri.org/publication/greenhouse-gas-protocol-corporate-accounting-and-reporting-standard-revised-edition.
2. *Building Solutions to Climate Change.* Pew Center on Global Climate Change; Arlington, Virginia (November 2006). Available at www.pewclimate.org/publications.
3. D. Roodman and N. Lenssen, "A Building Evolution: How Ecology and Health Concerns Are Transforming Construction," World Watch Paper 124, March 1995.
4. Leadership in Energy and Environmental Design (LEED) certification from the U.S. Green Building Council. See http://usgbc.org.
5. Amanjeet Singh, Matt Syal, Sue C. Grady, and Sinem Korkmaz, "Effects of Green Buildings on Employee Health and Productivity," *American Journal of Public Health*, July 15, 2010.
6. Marie Puybaraud, "Generation Y and the Workplace Annual Report 2010," Johnson Controls; Milwaukee, Wisconsin (2010).
7. BOMA Commercial Lease Guide. Available at http://shop.boma.org/showItem.aspx?product=GL2008&session=BEC6D61261054335969BA2A39E0F7B47.
8. A Survey of the U.S. ESCO Industry: Market Growth and Development from 2000 to 2006. Principal authors: Nicole Hopper and Charles Goldman; Lawrence Berkeley National Laboratory, Donald Gilligan and Terry E. Singer; National Association of Energy Service Companies, Dave Birr; Synchronous Energy Solutions. Ernest Orlando; Lawrence Berkeley National Laboratory. May 2007.

CHAPTER 7

Six Sigma Sustainability Project Examples

The typical corporate Six Sigma program gets deployed through a portfolio of design, measure, analyze, improve, and control (DMAIC) projects. In fact, we've worked with companies that focus so much on projects that they can even lose sight of the business value that the program was introduced to create. At these companies, black belts are certified by number of projects, projects are initiated against almost any goal, and pet issues are cast in DMAIC terms just to get resources. We do not advocate this approach. However, there is something to be said for taking a project mentality to your business. By identifying finite goals, enforcing the discipline of a sufficient business case, and organizing resources into a plan with a beginning, middle, and end, much can be accomplished. The project charter is an invaluable tool for management decision making because it allows the comparison of resource proposals based on value creation and clarity of purpose and plan. Just keep in mind that although a project uses Six Sigma or is headlined with sustainability, this does not mean that it gets an automatic green light.

In this chapter we'll make some suggestions for how you can use a project mind-set within your company to drive action on your sustainability program. We review examples of Six Sigma sustainability project charters and project case studies in three areas: energy conservation, real estate consolidation, and improved green lease quality.

Selecting Projects

The key constructs for selecting Six Sigma projects that have maximum sustainability value are all reviewed in this book: establishing the business case, collaborative management for continuous improvement, transfer functions, performance measurement, and so on. In the beginning of your corporate sustainability program, you won't have much useful quantitative data; you can expect to use a fair amount of qualitative information when targeting your first projects. The projects themselves can be used to create or to enhance the measurement systems for your most important outcomes. (Of course, if you are heading into territory that is not well instrumented, expect projects to take a bit longer and/or to have a bit more risk thanks to the ambiguity that comes from qualitative and more subjective standards for separating good performance from bad.) Project selection generally includes these principles:

- Use your collaborative management team as a project-management body. Centralized management of the project portfolio can provide useful input on the business value of proposed projects, can accelerate resource allocation, and can coordinate projects addressing the same parts of the transfer function.
- Break apart projects that address too many issues, chartering multiple projects for parallel solutioning in a cluster or for sequential solutioning as a multigeneration plan. This is more efficient than addressing a broad set of issues in one project. (As evidence for whether the project has too much scope, look at the goal statements: Once the goal statements are written in measurable terms, if more than one or two metrics are needed, the scope is probably too large.)
- Address a measurable outcome that is tied to a good business case.
- Assign project leaders that own or are willing to own the resulting process.

As project charters are being proposed, ask yourself if the project makes logical sense. How does it support the vision of the company for becoming more sustainable? In many cases, sustainability improvement projects might be doing less of something; don't be afraid to use defect rates as measurements. Look at the transfer function for your sustainable business, and the overlap with your business value drivers: Where can you create a shareholder opportunity? Look for the simple combination of two questions: Which of these issues are the most important drivers of our environmental impact, and in which areas are we performing the most poorly? *Most important* means that the issues are in the overlap of your environmental and business-value driver circles (see Chapter 1).

Because we work with clients around the world who are interested in reducing their energy consumption, we thought it only fitting to introduce our project examples using this topic.

Example Project: Reducing Cost and Carbon Through Energy Efficiency in Office Buildings

This project was done for a high-tech client but would be just as applicable for a company in any industry that uses a lot of office space. As with any Six Sigma project, there was the initial need to prove a get-started business case. Senior managers suspected that there was a cost-reduction opportunity but needed more information to make the case to launch the project. This information collection is typical of a sort of "pre-define" phase, where existing data are used to generate interest in the specific project. Management has finite resources, so before allocating staff, managers wanted to compare project ideas in a variety of areas. What information is needed to convince management that this project makes sense as a resource allocation? The answer depends somewhat on the management style and culture of the company—some companies are more data driven than others. The simplest way to look at this example is this: Based on our experience, if no formal energy-management program is in place at the company, typical potential savings are 15 to 20 percent over three to five years. This company had no formal energy-management program. Some managers want a bit more information; this is where benchmarking comes in.

For a project such as this, the typical benchmark metric would be energy intensity—kilowatt hours per square foot of office space per year. There are a number of good sources of high-level energy benchmarks, such as the Commercial Buildings Energy Consumption Survey (CBECS) conducted by the U.S. Energy Information Administration.[1] Since building energy consumption is a function of a number of factors, such as climate zone, number of occupants, and heating/cooling systems, it is also a good practice in the United States to enter building performance and feature data into the U.S. Environmental Protection Agency's (EPA's) ENERGY STAR Portfolio Manager system.[2] At the basic level, office space in an average building will consume about 17 kWh per square foot, and a poorly run office building will consume 23 kWh or more per square foot per year. Since our client was a high-tech company, we had to be sensitive to the potential influence of computer-intensive space, such as computer labs and server closets, the consumption of which in many offices is lumped together as office space.

In this particular case, our client had access to energy-consumption data for seven of its facilities, showing a consumption range from 23.6 to 40.9 kWh/square foot/year (Figure 7-1). Since the company had no formal energy-management program, we estimated the three- to five-year savings to be 15 to 20 percent, or roughly $570,000 to $760,000, for these seven buildings alone. The company occupied over 100 facilities. Assuming consistent operating practices and the ability to reduce costs commensurate with reductions in consumption, extrapolated savings would be millions of dollars over three to five years. Although in some facilities there were computer-intensive spaces, management agreed that there was a sufficient business case for launching the project into a proper define phase.

As the energy-efficiency project went into the define phase, key roles were filled. Because the improvement work would require resources from

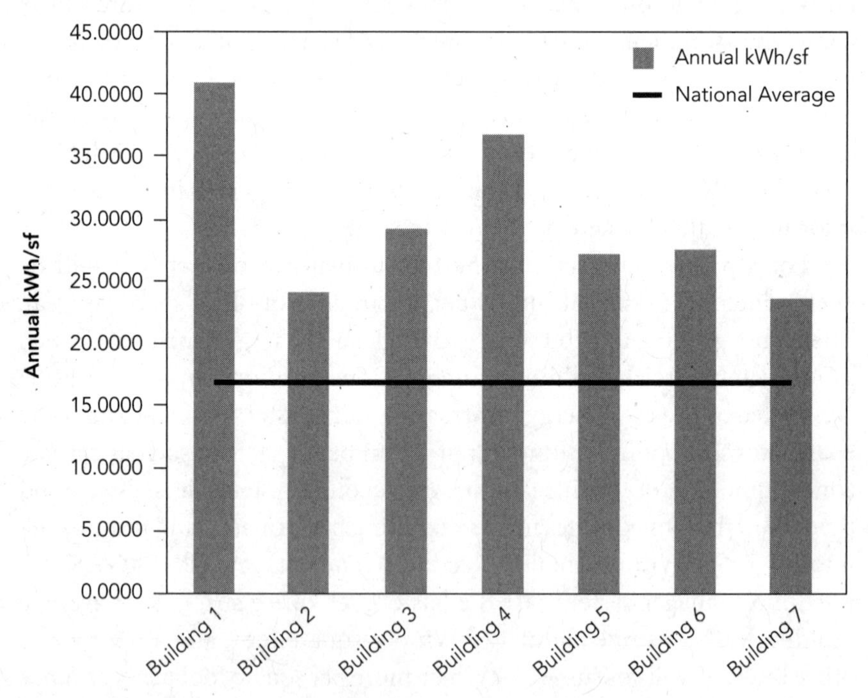

Figure 7-1 Annual kilowatt hours per square foot consumption.

the Facilities Management Department, it was decided that the role of project champion would switch from the director of sustainability (who was actually mostly interested in the carbon-footprint-reduction results of the project) to the director of Facilities Management. This was not as straightforward a decision as it sounded because this person was quite reluctant to devote resources to this effort because his already underresourced team was extremely busy with ongoing work. The compromise was made that third-party resources would be budgeted for and brought in to work on the project if the director/champion could assign a lead person from within his group. As much as possible, improvements to the building systems that would create the energy conservation would be made during the normal operation and maintenance of the facilities. Specific upgrade projects would be performed by outside vendors, if necessary. The project lead had some experience with Six Sigma and was partnered with a master black belt as a project coach.

There are a number of potential improvements to building systems energy efficiency that could be considered potential targets in a project such as this. As you'll recall from our illustration of the energy-consumption part of our sustainability transfer function, energy consumption in office buildings is a function of a few primary drivers, namely, such items as lighting, heating and cooling, operating hours, and controls technologies (and external factors such as weather). In a project for another client, we looked at each of these drivers across the portfolio at the macro level and found that aggressive steps had been taken in lighting and heating/cooling upgrades over the course of several years. In that other client environment, it was determined that only a broad-based recommissioning[3] program would result in any significant savings. So a cross-portfolio recommissioning program was initiated. In this particular project, as with other clients who are just getting started in energy management, it was decided to limit the scope of the improvements to the largest handful of buildings. Even though, technically speaking, this results in multiple project *y*'s, and we usually like to limit DMAIC projects to only one or two project *y*'s, it is easier to make many improvements to a few buildings. The project goal was set to reduce energy consumption by 5 percent within one year for the seven buildings in scope.

In the measure phase, the company looked at its data-collection system for energy data. For these particular buildings, collecting consumption data was not too difficult because the company paid the utility company directly, and the utility company invoices contained both consumption and cost data. In other facilities, data collection could get complex when energy use was invoiced by the landlord. The landlords also provided consumption data in many cases, but such data were included with the rent invoice instead of as a standalone document. The staff that paid the rent invoices did not parse out the energy costs and enter them into an energy-consumption tracking system, as did the accounts payable staff that paid the utility company invoices. There also were situations where rent invoices did not separately include a consumption number because it was just included in the rent. Essentially, the company had at least four scenarios for energy-consumption data in its facilities portfolio:

1. Direct data are fed from smart meters (approximately 5 percent of the portfolio).
2. Accounts payable clerks enter consumption data to the energy-tracking spreadsheets when paying utility company invoices (about 60 percent of the portfolio).
3. Facilities team enters consumption data into the energy-tracking spreadsheets when those data are provided on landlord rent invoices (about 30 percent of the portfolio).
4. Accurate energy-consumption data are not available because consumption is only an allocation within the rent invoice (about 5 percent of the portfolio).

These energy-data source scenarios would become important because the energy-conservation program would scale up to the entire facilities portfolio. For one thing, the operational definition of the *energy-consumption* metric would be affected by each scenario, where a different operator, a different format, and even a different calculation could be factors. Data-collection plans need to be very specific about what data to get and where to get them. But the team decided that overall energy-consumption reporting was not in the scope of its immediate project. And because of the inevitable repeatability and reproducibility issues that the company would experience as it endeavored to collect all energy data, the project team identified energy-data collection as a process in need of its own process-improvement project and moved on.

Separate from the data-collection process for ongoing energy consumption, the energy program would need to address measurement and verification of the energy-conservation improvements. For example, if upgrades are made to the building lighting system, or if behavioral changes are made, how will the facilities staff validate the amount of energy saved? The team did not address this at this point in the project but would refer to guidelines from the Efficiency Valuation Organization (EVO),[4] which publishes the *International Performance Measurement and Verification Protocol* for this purpose. These guidelines discuss how to apply a range of verification standards including engineering modeling/simulation options, weather normalization, metering, and the (less precise but cheaper) option to have the results stipulated based on standard calculations.

By the time the project team held its tollgate review, the team was already communicating with other Facilities Management staff to help highlight areas for improvement and to let people know that their buildings would need to be visited as part of an energy assessment to identify improvements. In conference calls with the managers of the in-scope facilities, the team learned that there probably were energy-conservation opportunities in abundance. The facilities managers had wanted to address these opportunities for some time but never got around to it. Based on information from the staff in the field, the project team confirmed to the project champion that the goal statement in the charter was still valid (in fact, they suspected that they could do a lot better than 5 percent even in the first year!).

In the analyze phase, the project team's focus was shifting from a review of energy-consumption patterns and measurement-system design issues to the specific drivers of energy demand in the in-scope facilities. The team contracted with an energy-assessment company to do a complete review of the facility. In one representative facility, it calculated the following profile of how electricity was used by the various systems in the building:

- Lighting: 26.62 percent
- Plug load: 16.64 percent
- Process electricity: 8.32 percent
- Heating, ventilation, and air conditioning (HVAC, fans and pumps): 31.61 percent
- Domestic hot water: 0.00 percent
- Cooling (summer extra): 13.05 percent
- Heating (winter extra): 3.77 percent

Although the building systems' electricity analysis conveys only "big *X*'s" (after all, moving consumption from one system to another is not really a conservation scheme), it is still useful for the facilities team to understand. Further analysis of each building system yielded a number of potential energy-conservation measures. As is typical in such assessments, the recommendations were categorized into two sets: low cost/no cost opportunities and capital improvement opportunities. Examples included

1. Reducing HVAC and lighting runtimes (which turned out to be an opportunity across the portfolio; the idea is to operate the HVAC and lighting equipment only when there are occupants in the buildings, taking into account daylight savings time and holidays)
2. Controlling the economizer with an enthalpy sensor instead of a dry-bulb sensor
3. Taking advantage of the two levels of lighting in the office spaces (e.g., using task lighting where possible)
4. Occupant and vendor energy awareness
5. Installation of second-generation T-8 fluorescent fixtures throughout the building
6. New direct digital controls for the HVAC equipment
7. Installation of window film on exposures where there are significant solar gains

Many of these improvement opportunities were found to exist across all seven in-scope buildings. By using the lessons learned from the on-site assessments, along with other known practices for building energy conservation, the team used the improve phase to develop assessment templates and ongoing process improvements (see Appendix). After testing out the solutions and verifying the results of some of them (which took a total of about three months), the team verified the expected energy-conservation and cost-reduction potential. Feeling that the company now had a viable portfolio of energy-conservation measures, and knowing that the company's buildings were consuming energy well above the rate of average-performing office buildings, the project champion adjusted the project goal to target 10 percent annual savings instead of 5 percent in order to achieve roughly 17 kWh/square foot by the end of a five-year period. The project champion also agreed to hire a full-time energy manager and contracted with a third-party support team. At the conclusion of the project, the team used the how-to information collected during implementation of each energy conservation measure; the information about savings in energy, cost, and carbon; and the company's new process maps to engage the rest of the Facilities Management team across the company. The team put in place a communications program for employees, as well as a training program for facilities staff, and started working on both performance improvements and making their measurement system more robust. Once the project champion was satisfied that the team was getting traction, the project was certified as passing the control phase, roughly nine months after the project was begun, but having captured a number of short-term wins along the way.

Example Project: Reducing the Environmental Impact of Company Office Space through Increased Office Space Agility

A consumer-products company client of ours was looking to take the next step into modern office space practices: more collaborative space, technology-enabled mobility, and higher densities without sacrificing access to private space. Over the past 10 or more years, companies such as Sun Microsystems, Applied Materials, Hewlett-Packard, and Unilever[5] have explored and deployed various strategies aimed at flexing the workplace to meet the needs of employees as they conduct their jobs. Changing desks in the workplace to sit near project team members, working from home or from "third places," shifting work hours, and virtual teaming create new ways of getting work done and new ways of accessing talented staff no matter where they live. And using workforce mobility as an advantage in rightsizing the office portfolio can have serious cost-reduction impacts while at the same time creating loyalty-driving employee benefits. In 2009, our client was expecting to invest over $50 million in office space to accommodate colocation of employees and business growth despite the fact that its existing office space was vacant 20 to 60 percent of any given day, the average office space allocation was about 250 square feet per worker, and the demand for employee mobility was increasing. The project champion and a manager who worked in the champion's organization felt that they could cut new office space investment costs by at least 30 percent in the first two years with a new "flex work" program. And the resulting cuts in energy use and carbon footprint were a nice sweetener to the financial logic. This fact- and data-based approach was enough to convince the management team that a business case existed to launch a project to design a new, more cost-effective workplace.

In the define phase, the champion worked across organizational lines to help the black belt assemble a team of the right subject-matter experts. Because the project was being approached from a real estate cost- and impact-reduction standpoint, the champion was from the Corporate Services Department, having a span of control over real estate as well as some other administrative services. The team had real estate expertise but also needed membership from the Human Resources (HR) Department (to represent the anticipated changes to management and work style) and from the Information Technology (IT) Department (to advise on application and other technical design elements). Addressing such a broad topic in a single project, not to mention the need to develop technology to support the new program, was a concern for the master black belt, who knew that these sorts of complexities can lengthen project duration significantly. The team felt that if off-the-shelf solutions could be used, it could get through a small-scale implementation within six to eight months. (*Post–project completion note:* It ended up taking almost a full year to get a real working prototype up and running.)

The project team was not very experienced with Six Sigma, which was apparent in the number of metrics it tried to address in the goal statement (nine). The team's list of goals was an attempt to address all the possible measurable benefits of the program:

1. Increase office space utilization by *X*
2. Reduce square feet of office per employee by *X*
3. Absorb headcount growth within the existing footprint (*X* employees)
4. Increase productivity by *X*
5. Increase employee satisfaction
6. Increase ratio of collaborative space to individual space
7. Increase ratio of shared to assigned space
8. Reduce carbon footprint
9. Reduce churn by *X*

After some discussion, the team determined that the most important goal for the project at this early phase was simply to reduce the cost of its office space by making the best use of existing office space, and therefore, the key metric became office space utilization. This meant that the team would need to increase the percentage of time that the office space was occupied or, conversely, reduce the percentage of time that the office was vacant. Some of the other potential metrics likely would emerge as customer requirements during the measure phase.

As opposed to a high-level process map as a deliverable in the define phase, the team used some images of the current office space layouts, assignments, and descriptive information to show the company the symptoms of this problem. As Figure 7-2 shows, there are underused conference rooms, underused office space, and other potential office design issues that could be resolved to improve access to daylight and produce a feeling of community (important drivers of employee satisfaction). The team also inventoried existing collaboration technologies and management practices. The company used collaboration tools for sharing files, but not very aggressively. And there were managers who were willing to let employees work from home if their job was a good fit. However, these practices were in place only unofficially and without any real business plan. In addition, even employees who worked from home periodically still had assigned space in the office; as a consequence, the company didn't really benefit from this arrangement.

Ultimately, the define phase tollgate decision was based on cost reduction for the real estate portfolio over time as well as on cost avoidance for a particular location, where one business unit was planning a 150-employee expansion and would be able to accommodate these new employees without acquiring new office space. With an average of 250 square feet per person and rents in that market at about $30 per square foot, the company would avoid roughly $1.1 million in gross new office rent by absorbing the new employees into existing space (not including retrofit costs). Employee satisfaction and containing the environmental impact of operations also were noted as opportunities for program benefits.

In the measure phase, the team identified the key customer groups as employees in general and managers. The real estate group would act as a proxy for shareholder interests in cost controls. (At other companies, such as a financial institution, the IT organization has strong requirements for data security. This issue existed for this company as well but was not regulated and did not come into play until later in the program design process.) The team ended up with the customer needs and related measures shown in Figure 7-3.

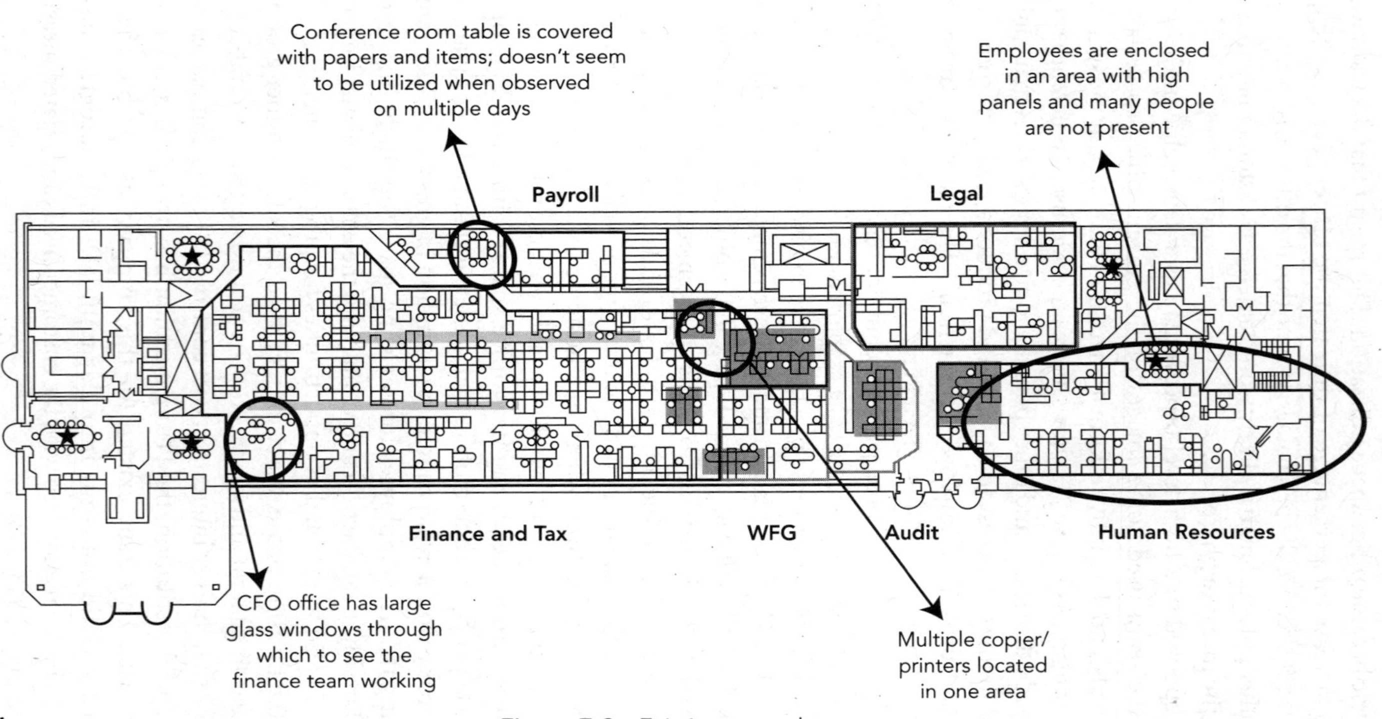

Figure 7-2 Existing space layout.

Customer	Need	Measures
Employees	Must have: Space to do my work when I need it. One-dimensional needs: Flexibility to work when and where I want. Nice to have: Mobility technologies (ultimately, this became a one-dimensional need.)	Utilization rates (if upper limits were exceeded, the assumption would be that employees did not have enough space). Satisfaction rates for feeling productive.
Managers	Must have: Confidence that employees are getting the job done; ability to find employees if they move around.	Satisfaction rates for employees accomplishing goals.
Real Estate	One-dimensional needs: Space that is dense enough to be more cost effective; space that is not vacant.	Space densities (square foot per person. Utilization rates.

Figure 7-3 Customer needs and measures.

By benchmarking with other companies, the team decided that an upper specification limit of 150 square feet per person for design density and a lower specification limit of 80 percent would be achievable for the offices it would put in the "flex work" program. The team did not have an existing measurement system for collecting utilization data, so it decided that for now it would have to rely on regular "bed checks." The team did bed checks at a handful of office sites to baseline current utilization. Figure 7-4 shows the results of the bed checks for one office.

The team noted that even during hours of peak vacancy, the office used a lot of energy to heat empty space, computers often were still powered on, and janitorial staff cleaned the space more or less the same way (with the same chemicals) regardless of use. The team was focused on the goal of increasing office space density but knew that it had to do this in smart ways, namely, without causing huge dissatisfaction and work disruption (simply putting smaller cubicles and offices in place without any other tools likely would backfire). The team needed to maintain or improve "workabilities" while at the same time increasing densities, which it knew would be seen by

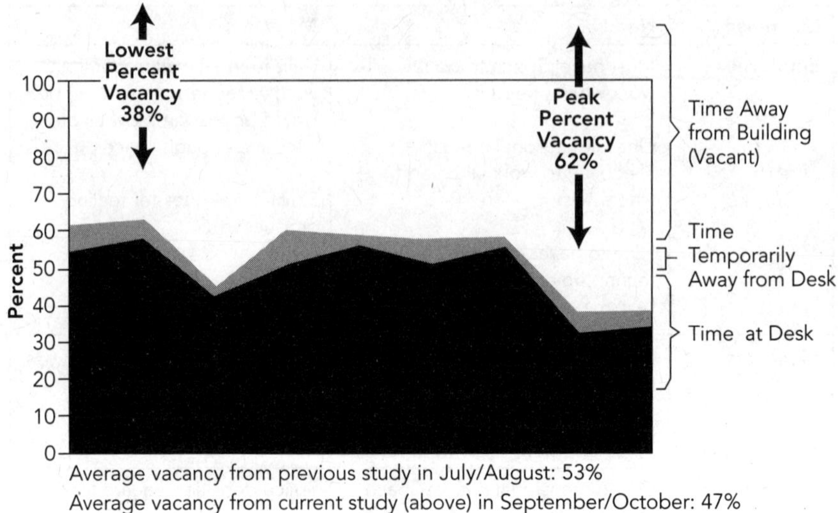

Figure 7-4 Utilization rates for an example office.

employees as a service takeaway if done poorly. The employee and manager satisfaction surveys would be important checkpoints.

In the analyze phase, the team needed to start creating a high-level design of its program. Team members wanted to know more about current practices that could be enhanced with more flexible workspace, and they wanted to look for clues for expanding employees' abilities to get their jobs done from multiple locations. The team used interviews and observations to answer two key questions:

1. When employees weren't in the office during work hours, where did they work?
2. When employees were in the office, how were they working?

The results showed that employees worked in a few predictable patterns (Figures 7-5 and 7-6).

The project team used this sort of where/how matrix as a brainstorming tool for design concepts. The team looked at how each need in the left-hand column might be satisfied under the condition at the top of each column (Figure 7-7).

At this point in the project, the design team knew that it was getting close to scoping the detailed design. Before moving on to that, however, the

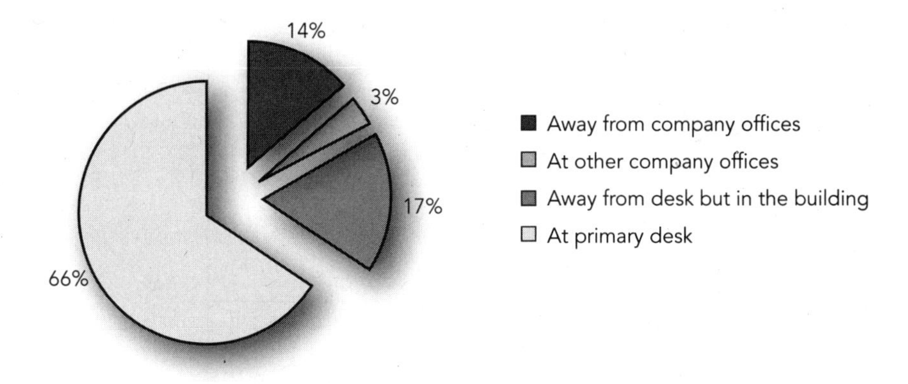

Figure 7-5 Where employees work today.

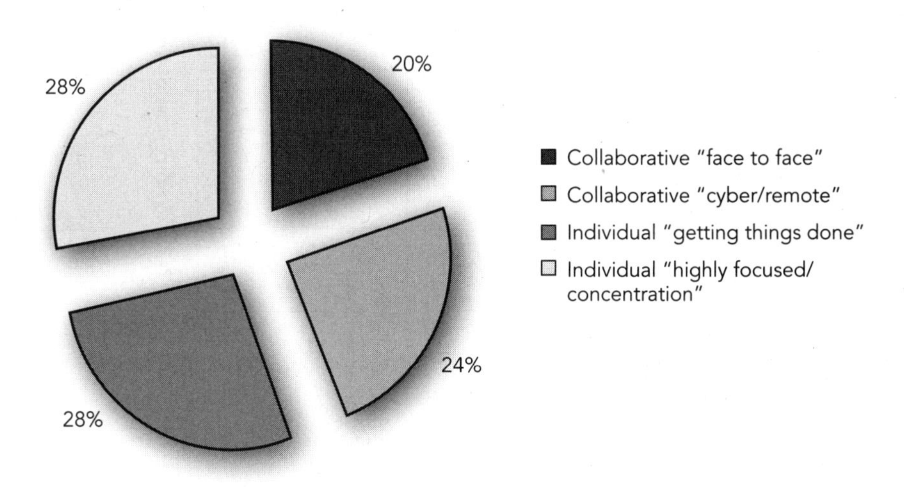

Figure 7-6 How employees work today.

team went back to employees and managers to get feedback on the design concepts. They also checked with stakeholders in HR, IT, and real estate. They found that managers still seemed reluctant to agree to the changes but committed to going along because the financial benefits were compelling and because some very senior executives agreed that they would participate in the new flexible space environment. The sales team that comprised a large part of the target offices seemed relatively indifferent, except that team

<table>
<tr><th></th><th>Away from Company Offices</th><th>At Other Company Offices</th><th>Away from Desk but in Building</th><th>At Primary Desk</th></tr>
<tr><td>Collaborate Face to Face</td><td></td><td></td><td rowspan="2">Could create more suitable collaboration space; need to make sure colleagues can find each other easily through registration system and "follow me" phone numbers.</td><td rowspan="4">Minimize need for assigned desk by creating other space types and the ability to reserve them when needed. Personal storage would be in central filing cabinets and in movable filing pedestals.</td></tr>
<tr><td>Collaborate Remotely</td><td>Need would increase; provide solid technolgy.

Allow and encourage more work-from-home for suitable job types.</td><td></td></tr>
<tr><td>Individual: Get Things Done</td><td colspan="2" rowspan="2">Need laptops with strong connectivity; company-paid Internet access; ability to access intranet when out of the office.</td><td>Smaller open plan/cubicles would work fine.</td></tr>
<tr><td>Individual: Highly Focused</td><td>Could create personal focus space available on-demand.</td></tr>
</table>

Figure 7-7 Where/how matrix.

members did not want to have to go through the transition of cleaning out their offices and learning new tools that would "slow down" their client-development activities. The design team started to realize that it was going to have to throw in some "delighters" in order to get more people onboard. These included upgrades to the space design standards so that the space would be "nicer," policies that (only) people who gave up assigned desks to go mobile and work remotely three or more days a week would have company-paid high-speed Internet access, and early adopters would get new laptops.

In the design phase, the team worked to detail the most important program elements and looked for performance risks. Team members also started documenting the program guidelines to create a program handbook. The key design decisions are listed below.

- Human factors:
  - Refresh management training on "managing by objectives" so that managers would not feel that employees who were not in the office were not working.
  - Use a screening process to determine whether employees were in jobs that were suitable for assigned desks or for remote work and to what degree.
  - Develop training for employees on remote collaboration topics.
  - Define policies for reimbursements for Internet technology and home office equipment.
- Technology:
  - Put in place "follow me" telephone systems so that employees could answer calls to their work number from a variety of programmable locations.
  - Make sure that software for virtual private networks (VPN), online document storage, and remote collaboration were on current versions and robust enough for increased use.
  - Create a space-reservation system and an employee-locator capability and set up technology support plans for remote workers.
- Real estate:
  - Update space design standards for the new densities, including furniture systems and upgraded furnishings.
  - Purchase centralized and portable storage units for employees.

- Design floor layouts with a combination of high-density open plan, collaborative spaces (conference rooms and soft seating), and heads-down private work areas.

As the project moved forward from design to verify, the project team built out a demonstration space at one of the existing offices to show employees what the new space would look like. As employees toured the new space, they were introduced to the new working concepts and benefits. The offices were rebuilt to accommodate the higher densities and the 150 new employees in the first office, which served as a pilot program. As it turned out, the company still acquired a bit of new space in that same building, which it used first as "swing space" and then to accommodate the consolidation of another office. In the end, the pilot space was showing utilization above the spec limits and also was more efficient than other offices of comparable size, achieving a number of new benefits such as electricity consumption reduced by 20 percent (roughly $20,000 per year) and water use reduced by 16 percent.

The project team leader became the overall program owner (and managed a steering committee with participation from HR, IT, an executive representative, and an employee representative). As the new system came online and employees got used to it, the space performed as intended. The program leader regularly reviewed the program scorecard—showing both utilization and energy and carbon savings—with stakeholders as part of a robust change-management program designed to get additional offices signed up to become flexible. Participation in the flexible work program was considered a green-team activity because it reduced an overall environmental impact by requiring the company to have less office space, reduced employee commutes, and contributed to stronger work/life balance.

Example Project: Greening Leased Space

Our client, a financial services company, wanted to put in place a program management office (PMO) that ultimately would drive cost savings across its portfolio of leased offices. The overall program objective was to reduce utility costs by 5 percent, improve the greenness of office operations, and reduce defects in the energy/carbon accounting process. Our primary scope included the processes necessary to reengineer the utility-payment and data-collection service model. The project involved redefining responsibilities among those who pay rents, those who administer the portfolio of leases, and those who manage the data. We used a DMAIC approach to tackle the project successfully.

Define

The first step in the define phase was to develop the business case; from there, we were able to draft the scope of the project and set goals. We determined that the best project-management approach would include specific speed and quantity goals against which the client could measure our efforts. We made sure that the project goals were both measurable (i.e., we could easily track performance using a baseline and a target) and realistic. It is important to have enough objectives to be meaningful but not so many that you cannot keep track of them all. For this project, we had a total of six measurable project goals:

- Number of leases abstracted
- Number of leases with calibrated utility operating-expense calculations
- Number of defects reviewed and tracked in our spreadsheet
- Creation of landlord utility charge/payment processes (i.e., field guides)
- Creation of accurate calculations of utility operating expenses for each lease
- Accurate identification of subleases for all sites, with utility cost accounting captured

Although cost savings was the ultimate driver of this project, it is important to note that savings were not one of our project success measurements. While savings are measureable, we could not realistically determine a savings goal at the outset of the project. Until we actually collected and analyzed the data, we could not know whether we would uncover savings in the client's portfolio.

Once we had established the business case, project scope, and goals, we were able to put together a project charter. This charter became an important document that we revisited regularly throughout the course of the project. In the charter, we identified the key stakeholders, whom we gathered early on in the project to conduct a stakeholder alignment workshop, during which we captured the voice of the customer. We heard from all parties about their hopes and concerns so that going forward we could address them all to ensure client satisfaction.

A final step in this phase was to document process maps for all the key processes affecting our project (e.g., utility-bill payment). This exercise allowed us not only to document the current state but also to engage in invited discussions of what the future state should look like (including redefined responsibilities and reporting structures).

Measure

With the help of key stakeholders, we identified a total of 54 leases that we would review. The criteria used for site selection included

- Size of the site
- Size of the utility spend at the site
- Lease structure
- Lease start date
- Lease end date
- "Tribal," or intimate, knowledge of the site (e.g., knowledge of other tenants in the building and their plans for lease renewal, lease termination, and so on)

Finally, we came up with a number of defects (21 total) that we believed might lead to cost savings if they existed in the client's portfolio. The defects were separated into three categories:

- Data completeness
- Tenant utility performance
- Landlord practices

In addition to tracking the defects for each of the 54 leases, we needed a way to capture language from the relevant lease sections, as well as the utility data information that we would be gathering. Ultimately, we developed a spreadsheet template to house all these data, with the rows containing information such as general property statistics, explicit lease sections, list of defects, and so on and a column for each of the leases.

We then went through the tedious process of reading leases and utility bills and then inputting the pertinent data into our master spreadsheet.

Analyze

Once we had all the necessary data in our spreadsheet, we could begin the analysis process. We methodically analyzed the information for each of the buildings against our 21 defects, assigning an X if the defect existed and a 0 if it did not exist.

Our next challenge was to determine what savings might be attached to existing defects. Using the criteria for whether or not a defect existed, we came up with a formula for calculating the savings. For example, the defect "landlord charge for energy is higher than market" existed if the landlord rate for electric consumption was more than 120 percent of the market rate. The formula for annual savings for this defect was

Annual savings = 50 percent × (landlord rate – market rate) × (annual electric consumption)

Using our information on potential savings, we created a matrix to depict the savings, listing the buildings in the rows and the defects in the columns. With a quick glance at this matrix, we could see the total potential savings for each building, as well as the total potential savings for each defect. We then could begin to prioritize which defects might have the biggest impact.

Improve

After we determined the projected savings, we realized that even by negating each of the defects, we would not likely realize 100 percent of the savings potential. Therefore, we established a probability adjustment methodology for each defect. For most defects, it was as simple as applying a straight percentage (e.g., 20 percent probability that the savings for the defect "Landlord charge for energy is higher than market" would be realized). For one particular defect ("Landlord could recommission building systems to reduce energy costs for tenants"), however, we needed to develop a more complex system.

We conducted an affinity analysis and determined that there were several factors that might influence whether the building would be a good candidate for recommissioning, including

1. Age of the building (or year of renovation)
2. Client's proportionate share in the building
3. Building efficiency [indicated by certifications, e.g., Leadership in Energy and Environmental Design (LEED)]
4. Expiration date of the lease
5. Terms of the lease (i.e., whether there was favorable language regarding passing realized savings onto the tenant)
6. Annual energy spend of the tenant in U.S. dollars per square foot

By denoting a rating for every factor (on a Likert scale of 1 to 5) and then weighting each factor according to impact (e.g., 15 versus 50 percent), we were able to assign an overall probability percentage to each building. We captured this information in a relationship matrix (i.e., quality function deployment).

Control

Once we were confident that we had measurable and quantifiable savings estimates, we were able to move into the implementation phase. We applied the Pareto principle, whereby approximately 80 percent of the savings comes from 20 percent of the buildings. Using this approach, we picked the top 8 buildings (of the original 54) that demonstrated the highest opportunities for cost savings and then implemented the recommended projects.

This process has proven to be repeatable and widely applicable not only to this particular client (for the remainder of its global portfolio) but also to other clients with large global portfolios.

Chapter Summary—Key Points

- Use your collaborative management team as the governing authority for identifying new projects based on performance improvement needs.
- Project champions need to have reasonable span of control over the resources required to execute the project. Project team leaders should own or be willing to own the resulting process improvements.
- Collecting preliminary data and comparing to industry benchmarks is an excellent "pre-Define" activity. Always start a project with the business case. For example, calculate a facilities portfolio energy

intensity (kWh per square foot per year) and compare to data from the Commercial Buildings Energy Consumption Survey (CBECS).

- Don't set too many goals for one project. In our "Office Space Agility" project, the team was trying to address office utilization, square feet per employee, employee satisfaction, employee productivity, amount of collaborative space, and four other goals all in one project.
- A "what/how matrix" can be a handy tool for identifying the features of a solution. One project team used this tool to identify how different employee work habits and work environments could be matched.
- Tie process metrics to financial goals for each project.
- Manage stakeholder communications as the project progresses, not just at the end.
- Project champions should reconfirm (or adjust) project improvement goals to reflect what the team learns as the project advances.

Notes

1. The Commercial Buildings Energy Consumption Survey (CBECS) is conducted by the U.S. Energy Information Administration every four years. The latest published findings as of this writing were from 2003 and were released in September 2008. Available at www.eia.gov/.
2. www.energystar.gov/index.cfm?c=evaluate_performance.bus_portfolio manager. Portfolio Manager helps you to track and assess energy and water consumption within individual buildings as well as across your entire building portfolio. Enter energy-consumption and cost data into your Portfolio Manager account to benchmark building energy performance, assess energy-management goals over time, and identify strategic opportunities for savings and recognition opportunities
3. *Recommissioning* refers to the process of reviewing current operations of building systems compared with how they were set up to operate when they were first put into use (commissioned). It is normal to expect building equipment performance to degrade and operating conditions to change over time. Recommissioning can achieve energy efficiency by eliminating the negative impacts of time.
4. www.evo-world.org/. According to its Web site, the mission of the Efficiency Valuation Organization (EVO) is to develop and promote the use of standardized protocols, methods, and tools to quantify and manage the performance risks and benefits associated with end-use energy-efficiency, renewable-energy, and water-efficiency business transactions. The EVO publishes the *International Performance Measurement and Verification Protocol,*

which is the widely recognized standard for measuring the impact of energy conservation measures.

5. The information about Unilever's Agile Working program is from the Unilever Web site: www.unileverusa.com/careers/campus_recruiting/worklifebalance/.

CHAPTER 8

Design for Six Sigma

Opportunities for companies to innovate are generated by market gaps. How these market gaps are identified—in other words, whether necessity is the mother of invention or vice versa—is less relevant for our immediate purposes than is the ability of the company to respond. As Six Sigma practitioners, we know that the best programs and products, because they last and grow via a continuous income of resources, must be designed from the outside (where the resources come from) in. It is in this space—between outside stimulus and internal response—that innovation can occur. Moreover, by putting a structure to that innovation, it can be channeled and monetized, and rework can be minimized.

The classic design tool for outside-in design is *quality function deployment* (QFD), also known as the "house of quality." The tool is so named because it is used to translate quality, as defined by customers, into functional requirements to be performed by the resulting product or service. Where a company sets out on the journey to design its sustainability program, we suggest that the house of quality makes an outstanding backbone for the myriad of business decisions about setting performance goals, managing handoffs across corporate functions, and making sure that department-level activities are in alignment with customer and shareholder expectations. Since a sustainability program includes a number of subprograms, the house of quality can be used again and again as a process for getting from the high-level corporate program definition to the next-level process and initiative designs.

For the purpose of this chapter, we will be using a simple version of the house of quality. In the Six Sigma industry, there are several variations of this tool that integrate quantitative targets, measurement correlations, and competitive benchmarking. In our version of the house of quality, we will maintain focus on the critical concepts rather than covering all possible features and uses of the tool. As in the transfer function we covered in Chapter 3, the house of quality emphasizes vertical relationships for how objectives can be attained. In addition, the translation of *whats* to *hows* maintains traceability between outcome measurements and activities that drive performance. (The house of quality is also referred to as a *what-how matrix.*)

In Figure 8-1 we show three different houses. House 1 is the "customer house," where, at the beginning of the design process, customer requirements are translated to measures. It is important to identify measurements that are sensitive to customer requirements because once the measures are identified, design focus shifts to process operation and task-level activities (in House 2 and House 3) to improve the measured performance. Since employees execute the activities, the house of quality ensures that the equation solves from the reverse direction as well. Employees perform their activities, contributing to overall process performance. In turn, the processes drive the measurements, and because the measurements were designed to be sensitive to customer expectations, we have ultimate confidence about how customers are feeling the impact of the performance of the overall sustainability program.

Below are the basic steps in executing a house of quality–based design process:

1. Identify and prioritize customer requirements.
2. Translate customer requirements into measurements that reflect these requirements.

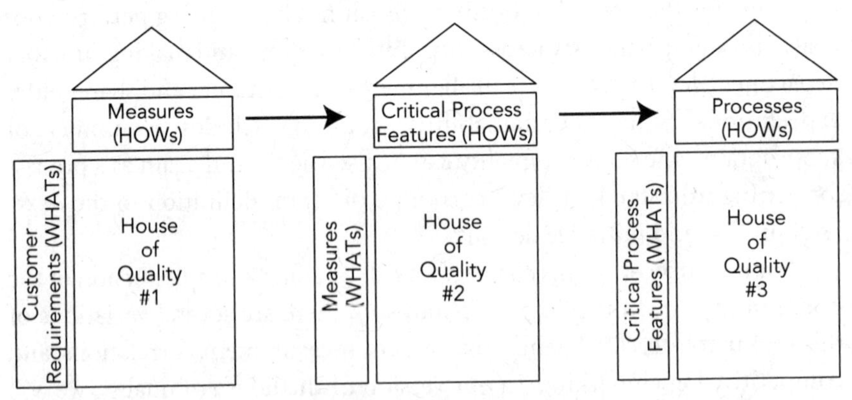

Figure 8-1 House of quality.

3. Set performance targets.
4. Identify critical process features required to achieve performance targets.
5. Design the processes that will meet the critical features.

As illustrated in Figure 8-2, the house of quality may be applied at three different levels within a company: at a high level to design the corporate sustainabilty program, at the core-process level for major systems such as energy management, or even at the project level when the goal is to design a product that is part of a larger process. In this chapter we'll show, first, how the house of quality applies by designing some of the major elements of the corporate sustainability program. Then we will show how it is used with the typical Six Sigma design project workflow using define, measure, analyze, design, and verify (DMADV).

Designing the Corporate Sustainability Program Using the House of Quality

Step 1: Identify and Prioritize Customer Requirements

The first step in any customer requirements definition process is to identify the customers of the program. You might recall from Chapter 2 that when establishing the business case for the sustainability program, Apex

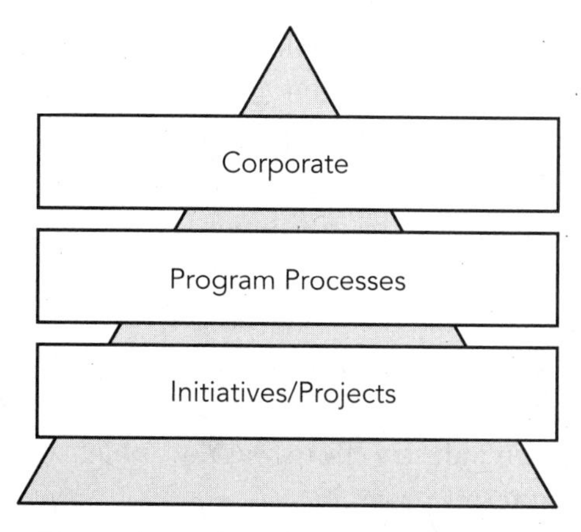

Figure 8-2 Three levels of application.

leadership looked at workforce interests, customer interests, and regulatory interests. In the workforce area, Apex was mostly interested in retaining its most talented employees and attracting new generations of talent. These are obviously common goals regardless of whether a company is pursuing sustainability or not. What Apex knew from employee surveys and other feedback mechanisms was that the environment was an area of interest, although not as highly ranked as compensation and supervisor relationship when it came to retention. However, in the talent-attraction area, working for a green company was an important differentiator, ranking closer to role, compensation, and title.

Apex also was interested in how to manage its carbon footprint because research from customers, competitors, and influential organizations described this as a priority.

For the purpose of this chapter, let's assume that Apex has two primary customer expectations:

- Appear environmentally responsible to employee-recruits
- Manage the company's carbon footprint

We suggest the use of a simple matrix for the purpose of organizing and analyzing voice of customer (VOC) information (Figure 8-3). In order to make the VOC measurable and actionable, we need specific information from the customer. The final result should be critical to the customer, measurable, and linked to action if specifications are not met. Through discussions with prospective company recruits from top colleges and further research with industry groups, the following matrix was completed for Apex in these two areas. To complete the "customer's house" in the house of quality, we need to identify good metrics for these customer requirements. Since the primary VOC is a relatively high level, we will use the key customer issues as the requirements to be measured. These are the specific issues we heard from customers that are important to them. It was a good thing we asked because showing that Apex is environmentally responsible could be done in a variety of ways—it was direct customer interviews that told us what customers meant by these statements. We also learned that participation in green industry efforts was not as high a priority as recycling and having a green teams program.

Voice of the Customer	Key Customer Issue
Show me how your company is environmentally responsible.	• Participate in green industry efforts • Recycle • Have an active green teams program
Manage your carbon footprint.	• Measure your carbon footprint • Set and achieve reduction goals • Report numbers transparently

Figure 8-3 Customer requirements matrix.

Step 2: Translate Customer Requirements into Measurements That Reflect Those Requirements

In order to identify the best performance metrics for each customer requirement, Apex made sure to involve experts in each area. In this way, the company was starting to develop buy-in from managers who would be responsible for providing data or for using those data or both. One common mistake in measurement design work is to choose metrics that are only loosely related to the customer requirement or even metrics that are not capable of showing progress against the requirement. For example, although recycling can reduce costs, using cost reduction as a measurement of a recycling program is not very effective. Cost reduction may be achieved in multiple ways. The key test is to focus only on the proposed metric and to ask the question, "If this metric were going up and *that's all I knew*, would I reasonably conclude that the customer requirement was being met?" If the answer is that the metric could be going up for other reasons, then you need a new metric. (See Chapter 4 for more information on measurement systems.)

After considering a number of possible metrics, Apex determined that the metrics shown in Figure 8-4 made the most sense for completion of the customer's house of quality.

Note that in this example we worked with the business to identify the specific level 2 requirement: "Show me." One could argue that this is not really a customer expectation so much as a business expectation—in other words, the employee being recruited is asking the company to show him or her, but the business recognizes that it won't get very much benefit from

First Level	Second Level	Percent of job candidates who are aware of recycling and employee program	Percent waste diversion rates	Number of employees in green teams	Number of metric tonnes of CO_2	Yes or No: Are carbon reduction goals set?	Yes or No: Were recent years' goals met?	Yes or No: Are footprint and goals published annually?
Show me how your company is environmentally responsible.	"Show me."	◉						
	Recycle.		◉					
	Have an active green teams program.			◉				
	Participate in green industry efforts.							
Manage your carbon footprint.	Measure your carbon footprint.				◉			
	Set and achieve reduction goals.					◉	◉	
	Report numbers transparently.							◉

● Strongly sensitive metric
◉ Moderately sensitive metric
○ Weakly sensitive metric

Figure 8-4 House of quality 1.

the green program as a recruiting tool if it does not make sure that recruits are aware of what the company is doing.

Step 3: Set Performance Targets

At this point in the process, it is important to assess your knowledge about what performance levels are important to your customers, as well as what performance levels your competition is achieving. If your customers expect a 50 percent waste-diversion rate but your competitors are achieving 80 percent, will customer expectations quickly rise closer to 80 percent? If your customers expect 50 percent diversion rates and your competitors are achieving only 30 percent, then there is evidence for a potential competitive advantage for your company. After obtaining customer input and competitor benchmarking data, work with the staff that will be managing performance in this area to continue understanding any business constraints. When you set your targets, consider these questions:

- Will this satisfy the customer?
- Will it make the company more competitive? Can we surpass competitor performance to achieve customer delight?
- Do we want to be leaders in our industry in this area? Leaders in all industries?
- Are these targets consistent with corporate goals? (For example, some companies do not publicly disclose operations data that would include carbon-footprint information.)
- What is our capability to achieve this measure?

Since this was a new program, the company needed to use information about competitors or other corporate programs of its own to help inform target setting. For example, Apex knows that another issue that is important to prospective employees is workforce diversity. Apex has had a diversity program for years and communicates this to prospective employees. Just as in environmental sustainability, Apex can't anticipate ahead of time which specific recruits will care most about diversity, but Apex knows from postrecruiting surveys that 85 percent of recruits know that it has a workforce diversity program. Apex also knows that with this 85 percent performance level, no prospects have refused a job offer for reasons related to diversity. This performance level seemed reasonable to the business, and there was evidence that it would suffice for recruiting purposes.

Additionally, this information implies that there are a couple of other potential metrics that a company could use in this area. If the real business issue (voice of business as opposed to voice of customer) is making sure that no potential employee turns down a job offer because he or she perceives the company to be unfriendly to the environment, at least two other potential results metrics could be designed to address this issue directly:

- Percentage of job offers declined where "lack of a sufficient environmental sustainability program" was identified as a decision factor
- Mean time (days, weeks, etc.) between job offers declined where "lack of a sufficient environmental sustainability program" was identified as a factor

The company could consider "percent awareness" as a leading indicator to "percent declined." (A corollary cause-effect pair of metrics also could apply for retention of existing employees.) Decisions about what and how many metrics the company wants to use are usually a function of the cost to monitor different processes and whether the data could be collected passively as opposed to via manual requests such as surveys. From a Six Sigma standpoint, more data are often better, but they are also often more expensive.

The bottom row of Figure 8-5 shows the addition of performance targets to the House of Quality.

Step 4: Identify Critical Process Features Required to Achieve Performance Targets

Let's focus on one category of customer requirements at a time, using the "first level" list as the organizing framework for our discussion. For the purpose of this book, we'll use narrative to describe the process of pulling together the right information. In a company environment, this work likely would be completed in a workshop format with lots of brainstorming to keep the discussion and output more visual. By focusing on process features before doing any process design, we start to sketch a conceptual design more quickly than we would if we were working on complete process flows. The resulting house of quality, after this step, becomes a design document for staff who can fill in the blanks for complete process documentation.

Addressing the first-level requirement—"Show me how your company is environmentally responsible"—our discussion focuses on the features our process(es) will need to have. The design team brainstorms process features based on its subject-matter expertise and knowledge of other relevant business processes.

Brainstormed process features for "show me" requirements include

First Level	Second Level	Percent of job candidates who are aware of recycling and employee program	Percent waste diversion rates	Number of employees in green teams	Number of metric tonnes of CO_2	Yes or No: Are carbon reduction goals set?	Yes or No: Were recent years' goals met?	Yes or No: Are footprint and goals published annually?
Show me how your company is environmentally responsible.	"Show me."	⦿						
	Recycle.		⦿					
	Have an active green teams program.			⦿				
	Participate in green industry efforts.							
Manage your carbon footprint.	Measure your carbon footprint.				⦿			
	Set and achieve reduction goals.					⦿	⦿	
	Report numbers transparently.							⦿
	Targets	85%	50%	500	900k	Yes	Yes	Yes

● Strongly sensitive metric
⦿ Moderately sensitive metric
○ Weakly sensitive metric

Figure 8-5 House of quality 2.

▲ In order to show the employee recruit that the company is environmentally responsible, the company will need a communications mechanism that reaches the prospective employee during the recruiting process. Communications becomes a process feature, as will the ability to aggregate basic information about the other requirements for recycling and green teams.

- The company also will need to communicate about something. So a recycling program of some kind will need to be put in place. It is unlikely that very many current or prospective employees will be technical experts on recycling. Therefore, in this particular case, a basic program is probably sufficient. That said, there may be other customers or key stakeholders who expect a more sophisticated program (e.g., for cost-reduction purposes or for more aggressive environmental benefits). The general employee population will care most about recycling of materials that they see every day (e.g., office supplies such as paper). But it wouldn't make sense to design two completely separate recycling processes (one for prospects and one for other expert stakeholders), so we would recommend that the design approach address all the recycling requirements that can be logically included at one time.
- Many of the participants in the brainstorming session mentioned that recycling is also about reduce and reuse. They felt that it wasn't good enough to just recycle more waste but wanted to reduce waste to begin with. Since our exercise focused primarily on office paper use, the team felt strongly that a responsible program would include paper use reduction through strategies such as double-sided printing. Related to reuse, one of the workshop participants noted that using paper that has a higher level of recycled content should be part of the program.
- As any office dweller knows, an important element of a recycling program is the bins where people are expected to put their paper for pickup. Our design would assume some number of receptacles where paper users would put paper and where janitorial staff would collect the paper.
- For the green teams program, our design team determined that employees would find this to be of value only if they could get involved in the program no matter where they were located. So one design feature of the program became geographic independence.
- The green teams program naturally would bring together employees with interest in environmental issues and would be volunteer-based. And in addition to being asked to participate in broad corporate employee programs such as recycling and energy efficiency, the design team thought that the green teams also could act as a test group of new employee sustainability programs before those programs would be rolled out across the company.

In summary, then, the key process design features for this show me set of requirements are

- Mechanism for communicating to prospective employees
- Recycling program for paper waste
- Reduction of paper consumption
- Use of paper with recycled content
- Convenient recycling receptacles
- Green teams that employees could participate in no matter where they work
- Green teams for environment-minded employees
- Use of green teams as early adopters for corporate programs and as testers for company sustainability programs

Next, the design team also brainstorms process features for the second requirement—"Manage your carbon footprint"—based on its subject-matter expertise and knowledge of other relevant business processes.

Brainstormed process features for carbon-footprint requirements include

- The design team's first question was, "What really is a carbon footprint?" After a brief discussion process, the team learned about the World Resources Institute's (WRI's) *Greenhouse Gas (GHG) Protocol* as a key reference guide to carbon reporting. Thus one of the design features became compliance with the *GHG Protocol* because this is important to other stakeholder groups.
- The carbon accounting process would need an efficient tracking mechanism for energy consumption and would have to easily account for the different carbon emissions factors for all the areas in which Apex did business. The team recognized that the carbon-footprint management process would be closely tied to an energy-efficiency program, which the company did not currently have.
- There would need to be a governance process put in place so that senior company managers could make decisions about carbon-footprint reduction goals and how efforts to achieve those goals would be funded.
- Since the carbon (and energy) management program would not be a one-time event of just counting carbon emissions, the program would need to assess facilities for improvement opportunities and report

progress on approved initiatives. The company was not sure if it wanted to install renewable power sources such as solar panels but figured the program would need to address these issues.

- The design team was aware that many companies publish an annual sustainability report to the outside world. The team decided that it would need a similar program so that key stakeholders would know about the company's progress in this area.

In summary, the key process design features for this carbon footprint set of requirements are

- Achieve compliance with the *GHG Protocol.*
- Track energy consumption and assign emissions factors for all business locations.
- Develop a governance process for setting goals and approving project funding.
- Complete a facilities assessment.
- Report on project progress (and on overall program progress).
- Address renewable power sources.
- Published a sustainability report.

With this information, the design team could complete house of quality 2 (Figure 8-6). By rating each process feature as having a strong, moderate, or weak contribution, the design team can ensure that each measurement is covered adequately (by looking across each row) and recognize which are the highest-priority design features for overall customer satisfaction (the features that are strong contributors).

Step 5: Design the Processes That Will Meet the Critical Features

Let's look at the process design question for the "Show me how your company is environmentally responsible" section of our customer requirements and design the high-level process that meets these requirements. As explained earlier, this is partly a function of communicating to prospective employees and partly a function of having something worthwhile to communicate.

Because the company already has a process for engaging with prospective employees, the design team wanted to connect to that process

	Mechanism for communicating to prospective employees	Recycling program for paper waste	Reduction of paper consumption	Use of paper with recycled content	Convenient recycling receptacles	Green teams that employees could participate in no matter where they work	Green teams for environmentally minded employees	Use green teams as early adopters for corporate programs and as testers for company sustainability programs	Compliance with WRI GHG Protocol	Track energy consumption and assign emissions factors for all business locations	Governance process for setting goals and approving project funding	Facilities assessments	Ability to report on project progress (and on overall program progress)	Renewable power	Published sustainability report
Percent of job canditates who are aware of recycling and employee program	●														
Percent of waste diversion rates		⦿	○	○	⦿										
Number of employees in green teams						●	●	○							
Number of metric tonnes of CO_2									●	⦿				○	
Yes or No: Are carbon reduction goals set?											●				
Yes or No: Were recent years' goals met?												○	●		
Yes or No: Are footprint and goals published annually?															●

● Strongly contributing feature (9) ⦿ Moderately contributing feature (3) ○ Weakly contributing feature (1)

Figure 8-6 House of quality 2.

if at all possible. At a high level, the new employee process, described from the perspective of the prospective employee, was a four-phase process:

1. Research the new position.
2. Indicate interest in the new position to the hiring company.
3. Participate in the company's screening process.
4. Negotiate with the hiring manager.

In order to best communicate the company's sustainability program to prospective employees, the design team had to understand where prospective employees naturally would get information about the company. Prospective employees used two main sources of information while researching a new company and a new position—existing employees and the Internet (both the company's own Web site and independent news coverage about the company). The design team decided that it could achieve its goals if it could get information about the company's sustainability program into these existing channels. The design team figured that it could control a push of program summary information to get it posted to the company's official Web site, but getting employees and the news media to promote the company's sustainability program would require creating a pull based on how compelling the program would be. During the company screening process, recruiters interacted regularly with prospective employees. The design team wanted to make sure that company recruiters could convey information about the company's sustainability program to prospects during this high-touch phase of the hiring process. The team talked to colleagues about how all these processes worked and brainstormed the resulting design requirements. The expectations the team heard from the Human Resources Department and the Web-site management team were referred to as the "voice of the (new hire recruiting) process" and are summarized in Figure 8-7, along with the process changes desired by the design team.

As for what content to put into the communications process, we have already explained that the design team is focused on the two expectations that prospective employees have for the company: recycling (of office waste) and green teams. As you will recall from Chapter 3, we identified the high-level process for paper use in a company:

1. Buy paper.
2. Use paper for printing.
3. Dispose of paper.

Process	Process Feature	Voice of the Process	Process Changes
Research new position	• Read the company's own website • Read media coverage • Talk to employees at the company	• Sustainability information posted to company website should be brief and engaging • Employees and media would talk about interesting and compelling stories (work with HR and corporate communications)	• Include information about recycling and green teams in the careers section of the company website • Promote company sustainability progress to media
Indicate interest in new position to hiring company	• Apply online	• Online application could be used to identify prospects interested in sustainability	• Add a check box to online employment application to ask prospect if he/she cared about sustainability as a criteria
Participate in company screen process	• Work with recruiter	• Recruiters could put sustainability information in packets • Recruiters fill out forms after each candidate meeting	• Provide brief summary information about sustainability to recruiters for packets • Ask recruiters to check box on candidate form when discussing sustainability
Negotiate with hiring manager	• Work with hiring manager		

Figure 8-7 Identifying changes to the new hire recruiting process.

In order to design the paper recycling process, the design team created a matrix of the process design features to this existing paper-use process. The process features were translated into activities for each phase of the paper-use process (Figure 8-8).

Because the recycling process was to be deployed more formally around the world, the company assigned a global process owner. This person was passionate and knowledgeable about reducing waste. In addition, as a true process owner, this person was given control over a budget for the labor and materials necessary to make the program successful. Control for all waste hauling and janitorial budgets was centralized under this person, with a service-level agreement established among this person, the corporate facilities management organization, and the corporate IT organization.

Although we won't go into the details here, the design process would continue for the green teams program and for the carbon-management program. The design process expands as more subject-matter experts are consulted, and the decisions become more granular in nature. After several of the program elements were put in place, the sustainability team anticipated a risk in the communications and reporting area. A number of customer requirements had been identified in this area, but instead of having each process owner designing an independent reporting process, the program team chartered a generic design project to apply the DMADV framework to reporting.

Design and Innovation in Projects Using DMADV

Many of the project concepts discussed in define, measure, analyze, improve, and control (DMAIC) also apply to define, measure, analyze, design, and verify (DMADV) projects. Both project types require a champion, a black belt, a project charter, and a team of subject-matter experts. The formality of this project organization structure should not be viewed merely as a feature of Six Sigma programs; we view it as an improvement to how business is conducted in general. Too many managers view Six Sigma as a concept that is separate from the rest of their business, whereas the real benefits come when the focus on customer requirements: data-driven decision making; risk assessment; and critical inputs, processes, and outputs becomes part of the operating fabric of the company. Individual design tools can be implemented for a myriad of business problems. Often, however,

	(Paper use process) Paper recycling process			
Design Features	(Buy paper) Procure recycled paper	(Use paper for printing) Print only what is needed	(Dispose paper) Segregate, sell, and measure	Report on recycling
Mechanism for communicating to prospective employees			Trash haulers to weigh paper that goes to recycling centers; calculate diversion percentage	• Use haulers' data to report aggregation • Recycling manager collects data monthly and creates summary report
Recycling program for paper waste			Trash haulers segregate paper for taking to recycling centers	
Reduction of paper consumption		• Improve double-sided printing • Use tools to reduce printing of unneeded content • Make sure excess paper is available for use in office		Make sure employees know the diversion rate and tonnage data for their offices (communicate via green teams and web or Twitter)
Use of paper with recycled content	Put recycled content requirements into procurement process			
Convenient recycling receptacles			Use centralized receptacles near printers to reduce cost and improve consciousness	

Figure 8-8 Paper recycling process.

managers mistake the time-intensive nature of DMAIC or DMADV projects during the initial learning curve of implementing a new Six Sigma program for a hard-wired feature of all Six Sigma projects. To maintain DMAIC/DMADV as both an effective and efficient problem-solving toolkit, skilled practitioners never lose sight of the principles involved and never treat a DMAIC or DMADV project framework as a prescriptive checklist of tools each of which must be applied at a granular level in the same exact way for every project. Rigor should be applied as a function of complexity and risk in a business problem. In this chapter we will focus on the important concepts of a DMADV project so that you can see how it applies in sustainability.

A DMADV project follows a five-step workflow:

- *Define:* Define the project goals and customer (internal and external) deliverables.
- *Measure:* Measure and determine customer needs and specifications.
- *Analyze:* Analyze the process options to meet the customer needs.
- *Design:* Design (detailed) the process to meet the customer needs.
- *Verify:* Verify the design performance and ability to meet customer needs.

The purpose of the define phase is to identify the new product or service to be designed or the existing project or service to be redesigned. In many cases, we hear managers say that the process they are interested in does not exist, and therefore, DMADV must be applied. Be careful of this temptation—DMADV assumes that a process does not exist and therefore requires a lot more customer input to make up for the lack of data. This makes the DMADV project look simple at the beginning, but as more customer groups are put in scope, the VOC collection process can become overwhelming in subsequent phases, and the project team collapses under its own weight. What many managers are actually saying when they tell you that the process doesn't exist is that the process is not documented and that different practitioners perform it at the task level in different ways. This is not the same thing as the process not existing. If you do it today, then start with DMAIC. If you purchase electricity for consumption in your buildings, manage building systems or production processes that consume energy, and make decisions affecting the energy efficiency of your business, then you have an energy-management process. It might not be documented. It might not be done the same way by everyone in your organization. But you have one, and this means that we would suggest you start with the DMAIC framework—because it's probably going to be easier to execute with a more narrow focus on one project *Y* and a few *x*'s. In many cases, we see business projects that start off as DMAIC projects only to discover, after the measure or analyze phase, that the current process shows no capability to achieve performance within the customer-specified targets. This is okay. Work with your master black belt to switch gears to a full process redesign, and finish the project under the DMADV umbrella. However, if you have determined at the outset that you haven't got a process capable of meeting customer requirements in some area, feel free to launch your project as a DMADV project. Review the charter with your master black belt and project champions for these key points:

- ▲ The new product or service concept is identified.
- ▲ There is a written team charter including project rationale, business case, preliminary business goals, scope, design budget, milestones, and roles/responsibilities.
- ▲ There is a list of targeted customers and a preliminary analysis of their needs.
- ▲ Any important constraints to process or service design or to business implementation have been identified.
- ▲ There is inclusion of findings from an analysis of stakeholders and their needs—concepts that will help to influence key stakeholders.

In a sustainability program, we often see companies exploring expansion from energy/carbon initiatives to make progress in other areas such as waste and water. It is entirely possible that a company without a large water footprint hasn't put much thought into water efficiency in the past and might want to design a new process in this area. A services company might find that its drivers of water consumption are fairly limited compared with a manufacturing company. However, because these issues haven't been explored in the past, quick application of the DMADV framework may be the easiest way to proceed—especially if you have a copy of this book and can use the transfer function we've defined!

It is worth mentioning that the problem statement for our Apex project as we design a new reporting program is not that "there is no reporting process." Instead, the project charter should be developed around the *opportunity* to influence business value, and in this case, opportunity is both brand reputation and company culture. (DMADV project charters often have an opportunity statement instead of a problem statement such as you would find in a DMAIC project charter.) One of the benefits of both DMADV and DMAIC projects is that project team members often are working together for the first time, and staffing the project team roster will require management to make important decisions, such as, Which public relations professional will be assigned to cover the company's sustainability program? Getting the sustainability lead, the public relations lead, and the HR lead for talent acquisition into alignment about messaging and media for attracting new generations of talent will feel like a small victory even before the reporting program is designed.

The purpose of the measure phase of a DMADV project is to identify the needs of the customers for the new process or service. During this phase, the project team will plan and conduct the customer research necessary to understand the customer needs and requirements associated with the product or service. It is possible that this information already exists within the company. Or if the design challenge is relatively simple, it might suffice to use assumed customer requirements rather than to take the time to conduct original research on the topic. Using assumed customer requirements introduces risk to the project—namely, the risk that the design team might design something that customers don't really want. But there could be sufficient information in the company, in the media, in research reports, or from other sources to make the project champion feel comfortable with this tradeoff. The measure phase has the following key objectives:

- Appropriately identify and segment customers, and collect information relevant to understanding customer needs and requirements.
- Prioritize customer needs and requirements.
- Translate customer needs and requirements into measurable characteristics that are prioritized with the measures defined. The team should ensure that it knows what a service defect would look like according to customers. (The discussion with customers about expectations and defects also can lead to discoveries of customer delighters, which are features that would greatly enhance customer satisfaction even though the features were not expected by customers.)
- Determine targets and/or specification limits, possibly through consideration of competitor performance levels.

In our sustainability program, Apex had already documented that the key customers include other companies to which it sells services, as well as regulators. And from a workforce and culture standpoint, we are treating current and prospective employees as customers. Although the design team could expand the list of customers, it is acceptable, particularly in the early stages of a sustainability program, to focus on the critical few. In our house of quality exercise, we already documented the requirements, collected VOC, and set measurable performance targets for some customers. This work generated a list of customer requirements for showing that the company is environmentally responsible and is managing its carbon footprint.

In the measure phase of a DMADV project (likewise in DMAIC), it is not sufficient to identify the performance metrics. At this phase of the project, the design team needs to fully design the measurement system and ensure that the measurement system has the capability to track results for the business. Creating a scorecard, even if it is based on a manual process, is an important step for showing how the new process will be managed. Now that the design team knows its customer needs, the metrics, and the targets, a scorecard for the reporting process can be developed. Based on the metrics discussed earlier in this chapter, we would expect the scorecard for this process to look something like Figure 8-9. At this point in the project, we also would expect the team to develop the detailed data-collection templates, including operational definitions, of each metric.

Measure Phase: Operational Definitions

The design team has now defined "percent awareness among job candidates" in a way that can be reproduced consistently by any data collector across the company. To enable this metric, we inserted a new checkbox into the online job application where applicants could indicate whether the existence of a company sustainability program (including recycling and green teams) is important to them. The applicants' responses (whether or not they check the box) become part of the talent-acquisition database. Additionally, employment recruiters are expected to include information about the company's sustainability program in applicant recruiting packets *and* ask all applicants who checked the interest box whether they are aware of Apex's program. Consequently, the operational definition of this metric is now the

Sustainability Reporting Scorecard

Placeholder for monthly data and lower specification limit line at 85%

Percent awareness among job candidates

	On track for this year?	Met last year
Are carbon reduction goals set? (Target: Yes)	Yes	Yes
Were recent years' goals met? (Target: Yes)	Yes	Yes
Are footprint and goals published annually? (Target: Yes)	Yes	Yes

Data reporting status

Percent locations with up-to-date data

Energy	95%
Carbon	95%
Waste diversion rates	95%
Waste tonnage to landfill	95%
Number of green teams members	95%

Comments

This space to be used for comments about reporting performance issues, recent improvements, and other information that will help managers.

Comment on any data collection issues such as region by region completeness.

Figure 8-9 Sustainability reporting scorecard.

total number of applicants who answer "Yes" to the recruiter's question about awareness divided by the total number of applicants who checked the interested box during the application process, as recorded in a specific area of the recruiting database. These numbers are reported monthly.

The design team also had to implement a new data-collection system for recycling data. During the segregate, sell, and measure phase of the recycling process, the company will rely on its waste haulers to measure how much paper is recovered from office waste. The waste hauler then will fill out a spreadsheet that the company has designed with monthly numbers for the weight of the paper, the weight of trash going to landfill, and the dollar value of any of the paper that the hauler was able to recoup from selling the paper as a commodity. The diversion rate is calculated as total pounds of paper diverted divided by total pounds of waste that went to landfill.

Once the design team has articulated the data-collection plan, it should be working with stakeholders to ensure that data collection not only will function as designed but that any staff required to track data also will be formally expected to do so. In this particular case, the employment recruiters play the key role in tracking the awareness of prospective employees. And before any of those process changes can be made, the IT Department has to implement the system upgrades in the recruiting tools. Any Six Sigma project that depends on IT systems changes must account for IT processes and cycle times in its project schedule. Our design team has a manual process to put in place while it waits for the IT Department to execute the changes it has requested.

The purpose of the analyze phase is to come up with a high-level design for the new process. In order to do so, it is recommended that the design team consider multiple design options. This is one place where the Six Sigma process creates opportunities for innovation. At the very least, asking design teams to consider multiple design options for meeting each category of customer expectations, document the options considered, and report the options that were not part of the final recommendation lowers the probability that the DMADV process will be used to implement a design scheme that was conceived before the project started. It is good management practice in reviews of major programs to ask the team for information about what other options it considered but rejected. This concept also can be applied effectively to nonsustainability projects as a way to ask about making a product more environmentally friendly. (Do the product design reviews in your company ask explicitly whether the design team considered other, more environmentally friendly options?)

At Nike, for example, both sustainability and innovation initiatives roll up to a single vice president who reports to the CEO. Nike participates in industry programs for the express purpose of applying the wisdom of the crowd to tricky product-design questions. If multiple companies all contribute to solving a tricky environmental problem, for example, coming up with an ecofriendly composition for rubber used in shoes, the research and development cost of the individual company can be lowered, and taking a greener product to scale across an industry can be accelerated.

The critical points for consideration before the design team should move on from the analyze phase include

- Identification of alternative concepts for the new or redesigned product or service and associated production process
- Best-fit design concept(s) selected from alternatives
- Use of resources outside the company for design idea generation, for example, benchmarking with other companies, using models from other industries/processes to generate creative solutions, and considering whether third-party vendors have a solution that could be purchased
- High-level designs developed to include (as appropriate) product/service, production process, information systems, human systems, facility, equipment/tools, and materials/supplies and design requirements developed for detailed design work
- Conceptual/high-level design reviews conducted to gather and incorporate feedback from customers, stakeholders, and experts
- Product/service risk assessment

One of the areas where we see innovation is waste reduction. In fact, we prefer to focus on waste-reduction processes as opposed to recycling processes because waste reduction implies a broader boundary for action, a larger system of moving parts and players, and as a result, a larger set of target business opportunities. One client we worked with looked at its waste-management operation holistically and created a new solution that reduced costs much more dramatically than we typically see in recycling-only programs. This innovation consisted of a number of changes:

- Removing all waste receptacles from employee offices and putting all receptacles in central locations around each floor. Although employees had to get used to taking their trash to another place instead of just dropping it under their desks, this process did make employees more cognizant of the waste they created.
- Switching to daytime cleaning reduced energy use by not having to light the office complexes after all employees had gone home for the day.
- Consolidating janitorial services into a single national contract. With more scale, centralized waste and recycling repositories, and daytime

cleaning, the new contract enabled the company to then negotiate significant savings from its suppliers.

For our reporting project, the analyze phase generated some discussion among our design team about the best way to publicize the carbon-footprint and other data. Although customers were not demanding frequent reporting of these numbers, the design team discovered that by using an enterprise technology software platform instead of spreadsheets, it would not be that difficult to show energy and carbon data on the company Web site practically in real time. With the emphasis on transparency among investors and other stakeholders, companies have been innovating their reporting programs to get away from static annual printed reports. (Printing was the first thing to go!) Having a sophisticated reporting program, including mapping data to individual locations, using multimedia, integrating carbon data with financial data, and showing data cross-referenced between multiple reporting systems, has become a customer delighter. At Apex, the design team wanted to include qualitative sustainability information on the corporate Web site, and since it also wanted to use social media to involve employees around the country in its green teams program, it was an easy conclusion to include a live green teams Twitter feed right in the careers section of its corporate Web site.

The design team executed a high-level risk analysis at this point. The entire team stood back from the conceptual design of the reporting system, including roles and responsibilities, technology, and workflow, and asked, "What could go wrong?" The team was nervous about the roles played by both recruiters and waste haulers because both groups were critical not just to the measurement of their process but also to its actual execution. Recruiters were necessary to both assess the knowledge of recruits and distribute information to them. And waste haulers segregated trash and weighed it. Without process controls, these groups might not perform their duties properly. At worst, they might misrepresent the actual performance levels of the process. The design team would consider requiring some sort of certification for waste-hauling vendors, for which ISO might provide a good standard. And for recruiting, the team decided that performing periodic spot checks by asking for feedback from newly hired employees or reviewing postrecruiting survey data might suffice.

For our reporting project, the team was reaching a critical decision about how to handle the sustainability data. There were essentially three options for managing carbon-emissions data across the company:

1. Distribute spreadsheets around the world and ask staff to complete the spreadsheets and send the resulting data to corporate.
2. Purchase a carbon accounting software tool.
3. Develop a carbon accounting database internally.

The spreadsheet option was dismissed because the team felt that the company was too big. Although it sounded easy (and cheap) to send around a spreadsheet, this could not possibly be the long-term solution. Manual data entry and manual integration of multiple spreadsheets seemed to lead to problems down the road; moreover, a manual system would not allow for the kind of real-time performance reporting that the team hoped to implement. There are so many good software platforms on the market now for carbon accounting that this seemed the more attractive option. That said, the company does have internal staff who can build databases. But there was serious doubt that the long-term cost of building and maintaining an internal system would be any less than buying a software system at the beginning. In addition, although the company could develop an internal database, it was not in the software application business and felt that "the pros" would do a better job of updating its software for new features and requirements. Management decided it would purchase an enterprise-wide carbon accounting software platform from an industry leader.

The purpose of the design phase in a DMADV project is to develop the detailed design of the product, service, and associated production process. It is also to evaluate the capability of the proposed design and develop plans to pilot the new or redesigned product or service. Before a design team can advance beyond this phase, the team should show that it has

- ▲ Performed a detailed design of all appropriate design elements
- ▲ Evaluated the detailed design to predict performance against customer expectations, functional, and/or design requirements
- ▲ Conducted detailed design reviews to obtain and incorporate customer, stakeholder, and expert feedback
- ▲ Developed plans to control the quality and cost of the product/service (early planning for process management)

- ▲ Developed plans to build and test a pilot version of the production process

In the design phase, our reporting team needed to complete the most important deliverables of its project. This included such items as:

- ▲ A detailed data-collection plan
- ▲ Templates for the company's annual report of its carbon footprint (and other metrics)
- ▲ First drafts of the one-page, high-impact information it would distribute to prospective employees via the recruiting department
- ▲ Detailed roles and responsibilities documentation for all the people involved in its reporting process
- ▲ Message concepts and draft language that the corporate communications staff could react to as potential communications about sustainability to the outside world
- ▲ Functional requirements documentation for both the internal technology changes (for the recruiting system) and the technology it would purchase for enterprise carbon accounting

It was important that the design team not do all of this work in a vacuum. The team was sufficiently connected to colleagues from different departments as part of the solution design. Team members were communicating regularly with stakeholders in both waste hauling and in recruiting. But these colleagues were not the consumers of the information that would be generated by the sustainability program. The design team shared sample information with employees to show what it was going to be publishing. The team also shared the information about the reporting process and its anticipated results on brand and culture with senior managers, whose continued buy-in would be critical for the continued commitment of resources in the program, not just resources required to improve energy efficiency and other improvements to environmental impact but also resources for generating a quality sustainability report that, for many stakeholders, was all they would see or know about what Apex was doing. (One team member mentioned that having a good report, in real time or not, was as important as having a good photographer at her wedding.)

The final phase of a DMADV project is the verify phase, the purpose of which is to build and pilot a full-function, limited-scale version of the new or redesigned product/service. During this phase, the project team will perform necessary startup activities and transition the new or redesigned product/service to the process owners. The criteria for meeting the final requirements of this phase of the project typically include

- Satisfactory completion of a test of the new or redesigned process, with test results analyzed and any final learnings integrated back into the process
- A final decision by the project champion and master black belt to scale up the process to full production mode
- Definition of a clear implementation plan to achieve scale, a control plan for performance, and procedures and standards for production
- Completion of startup activities
- Evidence that the production version of the product or service has been produced and evaluated against customer expectations, business goals, and customer response
- Completed transitioning of production processes to process owners

For the final project review, the reporting process design team had a significant body of work to take to scale across the business. By starting with external customer requirements, the team knew with confidence how to design the reporting process that would build Apex brand reputation and an internal green culture. The team knew what information would be needed to support the decisions or evaluations that would be made by customers, by regulators, and by current and prospective employees.

The team also knew how to measure the success of the reporting process based not on its own criteria for accuracy and data completeness but on metrics that showed whether its message was getting through to the intended audiences. Each function within the sustainability program, from green teams to energy management, knew which pieces of information needed to be provided to the reporting process and when. (All the roles and responsibilities were tested in a conference room pilot program, where the different functional leads used note cards to represent their information and passed this information along the routes prescribed by the process until the data reached test staff from recruiting, the media, and key customers.)

Because of the long lead time in making systems changes and in purchasing the carbon accounting software, the team had already started using a manual process that approximated the same data and workflow.

Team members were being trained on this process and knew that it eventually would be automated. Data about the results were shared with all participants so that people could see how their information was being used.

By transitioning to a process owner (someone identified early on in the design process), the team felt that it could safely close the project and turn over the scale-up process to the owner. Individual team members continued to work in their respective departments to implement technology changes or work with third parties, such as waste haulers, to continuously monitor these new tasks. The process owner was already meeting monthly with managers from the sustainability program, facilities, corporate communications, and human resources to review the process scorecard and collect ongoing comments for future enhancements.

At the final project team meeting, the project champion distributed small awards to each team member and thanked him or her for taking on this task.

Chapter Summary—Key Points

- Design tools such as the house of quality are important for structuring a company response to market gaps and other needs for innovation. The house of quality (also known as a *what-how matrix*) can be used to design a corporate sustainability program, core processes, or individual products.
- The basic steps in executing a house of quality–based design process are
  - Identify and prioritize customer requirements.
  - Translate customer requirements into measurements that reflect these requirements.
  - Set performance targets.
  - Identify critical process features required to achieve performance targets.
  - Design the processes that will meet the critical features.
- We illustrate two examples of applying the house of quality: (1) using sustainability in designing an employee attraction program and (2) designing a process to manage a company's carbon footprint.
- Six Sigma design projects often use the DMADV approach:
  - *Define:* Define the project goals and customer (internal and external) deliverables.

- *Measure:* Measure and determine customer needs and specifications.
- *Analyze:* Analyze the process options to meet the customer needs.
- *Design:* Design (detailed) the process to meet the customer needs.
- *Verify:* Verify the design performance and ability to meet customer needs.

- Other key tools in the design process include the failure modes and effects analysis (FMEA), which can be used to identify and prioritize sustainability risks. The FMEA tool prioritizes risks based on severity of impact to the business/environment, likelihood that the risk will occur, and how easy or difficult it is to detect that risk in time to act.
- The important elements of a program scorecard include charts showing data over time, performance against critical feature requirements, data-completion information, and comments about reporting challenges, recent improvements, and other information to support management decisions.

CHAPTER 9

Stakeholder Management

Influencing the opinions and actions of groups that can have an impact on your company's short- and long-term sales and stock price is always tricky. When the topic is your company's performance relative to environmental sustainability and climate change, the number of stakeholders has grown dramatically over the past few years, as have the breadth of issues reviewed and the sophistication of the analysis. Many companies have shifted their goals for stakeholder engagement from risk management and talking only about the good things the company is doing to collaboration and transparency in discussing performance strengths and weaknesses.

Cisco summarized its stakeholder engagement goals in its 2010 Corporate Social Responsibility Report[1] as follows:

> Engaging with our stakeholders, the people and organizations that affect or are affected by our business, helps us align our business more closely to society's needs. Their input also helps us assess the issues that are most material to our business. Through these interactions, we aim to
>
> - Gain valuable information on external perceptions of Cisco
> - Obtain specialist insight from stakeholders with expertise in our industry on relevant CSR [corporate social responsibility] issues
> - Build ongoing relationships with key influencers
> - Update our stakeholders on our CSR efforts
>
> Engagement with CSR organizations also enables us to benchmark our performance against our competitors and peers, and annual feedback sessions on our CSR reporting help us

identify strengths and weaknesses in performance, disclosure, and readability.

These days, for a company to be considered an effective reporter of its own sustainability progress, stakeholders expect transparency in both performance and process. They want to know what's in the sausage, as well as how it's prepared—or, if you prefer, your company is expected to be transparent about how it's being transparent!

Consistent with Six Sigma principles of outside-in design, the best practices in sustainability programs include engagement with key stakeholders. The engagement process is cyclic and, thanks to social media, can be continuous. Your company will be expected to collect stakeholder commentary, assimilate input into your goal-setting process, and set expectations for future areas of progress. This is not to suggest that all stakeholders are of equal priority, nor should you expect that the stakeholders speaking the loudest should become, a priori, your highest priority. It also will be tempting to use all manner of media for communications with stakeholders. As with performance data, however, while more stakeholder engagement is better, it is also more expensive. The structure we propose in this chapter will help you to make proactive decisions about how to engage your most important internal and external stakeholders. As we go through each section, we will provide suggestions as input for your early decisions about which stakeholder groups to engage, how to do so, and how to structure the evolving dialog and influence strategies. In some ways, stakeholder engagement is easiest at the beginning of your sustainability journey because everything seems possible. Without a good set of practices and processes, however, your company may be led down a path that eventually becomes ineffective, expensive, and difficult to change.

In order to build our stakeholder management processes and practices, we once again look to the existing body of work from the Six Sigma community. In the Six Sigma framework, a stakeholder can be considered another variation of a customer. Even though many stakeholders don't actually "pay" for the services they receive, they hopefully derive benefit from the services delivered. For this reason, we will draw from the work of Six Sigma practitioners for insight into building practices and processes that are focused on better understanding customer needs and expectations and then converting those needs and expectations into program or process requirements.

Defining Stakeholder Management

At the risk of stating the obvious, a *stakeholder* is a person or organization that has a stake in the outcome of the program or process. That is, the stakeholder stands to gain or lose personal or organizational benefits from the process. Because stakeholders believe that they will either suffer loss or enjoy benefits, they will either support or oppose the process or program. Stakeholders will support or oppose a program based on their perceived needs and expectations in relation to the program. Their level of support or opposition will increase or decrease based on the gap between their perceived needs and their perception of what the program is actually delivering. *Stakeholder management*, therefore, is the set of processes and practices that enables one to deeply understand, monitor, and influence the perceived needs of various stakeholders and then build programs that will deliver on those needs or put in place strategies to positively influence the perceptions or mitigate the risks associated with opposition.

Understanding Stakeholder Needs

The science of understanding customer needs is another area in which a great deal of research has been conducted and much of the practices codified. Our intention here is not to summarize all those practices but rather to highlight the necessity of clearly understanding stakeholder needs and suggest basic practices that can be employed to gain some insight into those needs. We find it important to do so because, as we reviewed sustainability programs that have failed, a key failure mode has been neglecting or misunderstanding a key stakeholder or that stakeholder's specific need. It's a failure mode that can be avoided by employing some basic practices that will enable identification of your stakeholders and an understanding of their basic needs. We will go into identification of stakeholders that are unique to a sustainability initiative later in this chapter. Our purpose at this point is to explain some basic techniques for gaining insight into stakeholder needs. Happily, the most effective practices are simple to implement. Surprisingly, too many teams engaged in sustainability initiatives fail here. The basic practice, once your stakeholders have been identified, is to ask, listen, and then clarify. Teams often skip this step because they assume that they understand the needs, or they assume that the stakeholder's opposition will be immaterial. "How could anyone in their right mind be opposed to my green initiative?"

In Chapter 5, on change management and the power of teams, we introduced an approach to learning stakeholder expectations that is worth reintroducing here. A simple question posed to each stakeholder group will bring out insight into what the stakeholders expect from a sustainability initiative. Team members should brainstorm stakeholder response to the following fill-in-the-blanks statement: " In order to meet my expectations, this sustainability initiative must. . . . " Typical responses to this exercise could include those listed in Figure 9-1.

Converting Needs to Requirements

We can carry the preceding example further to demonstrate how this voice-of-the-stakeholder activity can be extended to enable the team to develop explicit needs and then convert those needs into process or program requirements. The team should review each of the stakeholder verbatim

Stakeholder	Response
Customers	"In order to meet my expectations, your organization's sustainability initiative must provide me with a product that does not harm the environment and a product that costs less than your competition."
Employees	"In order to meet my expectations, your organization's sustainability initiative must provide me with opportunities to directly impact the effect that my work has on the environment and allow me to have a voice in how we improve our environmental impacts."
Shareholders	"In order to meet my expectations, your organization's sustainability initiative must deliver a positive return on investment, while reducing the overall organizational risks associated with negative environmental impacts."

Figure 9-1 This basic voice-of-the-stakeholder exercise starts with team member brainstorming, which then is expanded and validated through group or individual interviews with the stakeholders, and finally, quantified through surveys.

quotes to determine what concrete need that stakeholder group is voicing. In this case, the needs can be drawn directly from the statements. In most cases, though, these statements require significant interpretation, debate, and further dialog with the stakeholder group. It is this iterative process with the stakeholder group that leads to clear agreement on needs and expectations and ultimately drives the various parties to positions that each party can support. Once the explicit needs are agreed on, the next step is to ask the question, "If this is the defined need, what specific requirement does this need place on my process or program?"

Often a single need can generate multiple requirements. The key to success in this activity is to develop program or process requirements that are concrete, specific, and measurable. The follow-up to this activity is to show the resulting *requirements document* to the stakeholder groups and give them an opportunity to debate and prioritize the relative importance of each requirement. The finalized document, validated by the stakeholders, demonstrates the clear linkage to stakeholder needs and provides a clear requirements framework that can help to drive the design of the overall sustainability initiative.

Once the team has documented stakeholder needs and process requirements, it is critical that team members develop a plan that will enable them to manage the stakeholders in ways that satisfy those needs and fulfill the requirements. The relationship-management plan is a practice that enables teams to understand who the key stakeholders are and how their needs/concerns will be addressed as the overall sustainability initiative moves forward (Figure 9-2).

In Figure 9-3, the teams have identified key individuals in each of their stakeholder groups. They then use the relationship review tool to begin to plan a strategy for positively influencing each of these stakeholders. The strategy for each is rooted in the stakeholder requirements developed in the previous exercise. The strategy is also shaped by four other factors:

- The type and level of influence that the stakeholder has with regard to the sustainability project or initiative
- The assignment of a key team member to own this relationship and to drive the strategy with this group or individual
- The strength of the relationship that the group or individual has with your organization or team

Stakeholder	Explicit Need	Requirement
Customers	"A product that does not harm the environment and a product that costs less than your competition."	Carbon neutral products Lower product costs
Employees	"Opportunities to directly impact the effect that my work has on the environment and allow me to have a voice in how we improve our environmental impacts."	Increased employee involvement Increased employee dialogue
Shareholders	"A positive return on investment, while reducing the overall organizational risks associated with negative environmental impacts."	ROI above 15% Reduced risks

Figure 9-2 Developing a relationship-management plan.

- The extent to which the group or individual is likely to support or oppose the team's project or initiative and the factors that are working for or against that support or opposition

Notice that the owner of the relationship is not always the company's chief sustainability officer. This is an important opportunity to integrate sustainability into the different functional areas of the business and make sure that managers who "speak the same language" are the ones who engage the stakeholder groups. You probably would not ask the chief financial officer (CFO) to own the relationship with environmental regulators. But the CFO and other leaders in the Investor Relations Department would be very effective relationship owners with institutional investors. Also keep in mind that although we tend to think of stakeholder groups as homogeneous institutions, they are, in fact, composed of individuals who each have their own style and perspective—your success engaging Greenpeace may be more a function of the rapport established between your individual relationship owner and the specific Greenpeace representative than a function of a company-to-institution formula. Every relationship owner must be involved in the analysis of the role of his or her stakeholder groups in the overall engagement program, as defined by the four factors mentioned earlier. In the next few pages we will detail these four factors that shape strategy and stakeholder roles.

Name	Stakeholder Group	Requirements	Decision Maker, Influencer, Champion, or Blocker?	Owner	Strength of Personal Relationship	Strategies/ Comments
Ms. Mary Smith	Customer	Carbon-free products and low product costs	Influencer/ Champion	VP of Marketing and Sales	Moderate	Publicize carbon reduction impacts that have occurred over past 12 months
Howard Jones	Employee	Increased employee involvement and employee dialogue	Blocker	Direct of Human Relations	Poor	Implement employee engagement and sustainability education program
Ron Brown	Shareholder	ROI above 15% and reduce overall risk	Decision Maker	VP of Investor Relations	Strong	Business case presentation for Board of Directors

Figure 9-3 Key individuals in the stakeholder group.

The Stakeholder's Type and Level of Influence on the Sustainability Project/Initiative

Because each stakeholder believes that he or she will either suffer loss or gain benefit from the sustainability initiative, stakeholders are likely to take certain actions that can either hurt or support the initiative. Based on certain factors, such as authority, position power, purchase power, and personal relationships, the actions they take will have a varying level of positive or negative impact on a sustainability project or initiative. For this reason, it is useful to categorize influencers so that specific strategies can be employed to improve the stakeholder's perception of benefit or reverse the stakeholder's negative perceptions in ways that will either leverage or neutralize the impact of his or her actions. Three commonly used categories for stakeholders are decision makers, champions, and blockers.

Decision Makers

The decision makers are usually the easiest to identify. Within the company, think *managers*, and outside the company, think *regulators*. In both cases, decision makers typically are the people with formal authority for the scope of your program's issues and initiatives. They have the ability to drive an initiative forward or stop it by providing people, money, or their own time. Because their decisions usually focus on where, when, and how to deploy a limited set of resources, they are in the business of reviewing various requests for resources, understanding the organizational tradeoffs between the various requests, and then deciding which initiatives will receive the resources. The most effective method for influencing this decision-making process is to provide a clear cost-benefit analysis. Creating a cost-benefit analysis was discussed in great detail as part of preparing the business case in Chapter 2. For our purposes here, it is important that the benefits and costs are delineated in a way that enables the decision maker to clearly understand the net gains and compare those net gains with other potential projects. When possible, benefits and costs should be quantified, and dollar values should be assigned. While the process for positively influencing a decision maker is fairly straightforward, the strategy for reversing negative perception can be much more difficult. Decision makers are often unwilling to disclose the reasons for their negative perceptions or their reasons for saying "No!"

to a project. When such a situation occurs, identify a supportive individual who has some influence with the decision maker to inquire and advocate on your behalf. Using information that the supportive individual gathers, you then can change the positioning of the initiative or revisit your cost-benefit analysis.

Champions

The individuals who are highly supportive of your sustainability project or initiative and have the willingness and ability to inquire, advocate, or remove barriers on your behalf are know as *champions*. They are useful because they can help you to overcome problems with decision makers, and they can help to neutralize or even turn around groups or individuals who are blocking an initiative from moving forward. An important strategy for managing champions is to proactively search out the individuals in positions of power and influence who stand the most to gain personally from the sustainability initiative. The most effective champions are rarely motivated by altruism alone. They are best motivated when they understand how their early support of the initiative will help them toward common objectives or build personal stature. A cautionary note here is to beware of false champions. False champions are motivated to support your project because they see personal gain, but they lack the time, resources, or personal power to actually provide the support and influence that you need to move your initiative forward.

Blockers

The last category of stakeholders is blockers. Groups or individuals who are blocking an initiative from moving forward are known as *blockers*. An interesting aspect of blockers is that they can be located just about anywhere and on any level in an organization. They can be part of the decision structure, or they can be in your user/employee base. Typical blocker tactics could include the simple withholding of support, negative communication campaigns, or launching counter projects. Blockers are often motivated out of a sense or perception of personal or positional loss.

An important strategy for managing blockers is to proactively identify who the blockers are likely to be and understand what will be driving their negative perceptions. You can then address those perceptions in the design of the campaign and in a series of change-management tactics.

Type of Influencer	Strategies for Positively Influencing	Strategies for Reversing Negative Perceptions
Decision Maker	Provide a cost versus benefit analysis	Identify a champion to inquire or advocate on your behalf with the decision maker
Champion	Proactively identify and cultivate the champion by understanding how they will personally gain	Remind the champion how they are likely to personally benefit from the initiative
Blocker	Understand the blocker's negative perceptions and design the initiative and the change management programs to address those perceptions	Use champions to address the blocker's concerns

Figure 9-4 Sample stakeholder influence strategies.

The mistake that teams often make is in their assumption that blockers don't exist or that the shear force of the initiative can overcome them. Such a rationale rarely works. It's important to identify and proactively manage the blockers (Figure 9-4).

Assignment of a Team Member to Own a Stakeholder Relationship and Drive the Strategy

Once the key stakeholders have been identified, requirements validated, and level of influence categorized, an *owner* should be assigned to manage the specific relationship with that stakeholder. Key to success in matching owners and stakeholders is to identify certain affinities between an owner and a stakeholder. The owner should have certain skills, knowledge, or traits that will be interesting or important to the stakeholder, which will enable the owner to exert influence over the stakeholder. Often positional power can be an important trait. That is, the owner should have a relative organizational position comparable with that of the stakeholder. For example, a first-line engineer is likely to have difficulty influencing a CEO, and conversely, a CEO may intimidate a first-line engineer and therefore not be able to positively influence that engineer. Another trait that seems to be important in the sustainability arena is academic background. Matching owners and stakeholders with levels of education and similar degrees improves the likelihood that positive influence will result.

The point here is that the assignment of owners to stakeholders should be a proactive, deliberate, thoughtful activity. Careful thought must be given to the likelihood that the match of the owner to the stakeholder will result in a positive influence on the stakeholder.

Once an owner has been assigned, the team should make a realistic and ongoing assessment of the strength of the relationship that each stakeholder group is likely to have with your sustainability initiative. *Relationship*, in this case, means the manner in which the stakeholder relates to your initiative. This activity is an attempt to assess the extent to which a stakeholder group views your initiative positively or negatively and how likely they are to take positive or negative action based on their views of your initiative. The goal is to identify or cultivate stakeholder relationships, where the stakeholder relates very positively to the initiative and is likely to take actions that will help the initiative to be successful. Equally important is to identify and reverse the views of stakeholder groups that relate negatively to the sustainability initiative and are likely to take actions that can impede the progress of your initiative.

A practice that can help you to assess the strength of the relationship and develop strategies to either counter negative perceptions or build positive perceptions is to rate each relationship along two dimensions—knowledge and rapport (Figure 9-5).

The knowledge dimension assesses the extent to which a stakeholder group or an individual accurately understands your initiative. Does the stakeholder group understand the drivers behind your initiative, and do they understand the positive outcomes or personal benefits likely to result from your initiative. Assessing where your stakeholders lie within the knowledge dimension is important because it helps you to plan your communication and education strategies. Focusing here will enable a team to determine how much education is necessary for this stakeholder group and what key knowledge gaps actually need to be filled. It also will help team members to understand whether or not an initiative is being explained or positioned in a way that is too technically complex for a particular stakeholder group to understand the drivers, benefits, or outcomes.

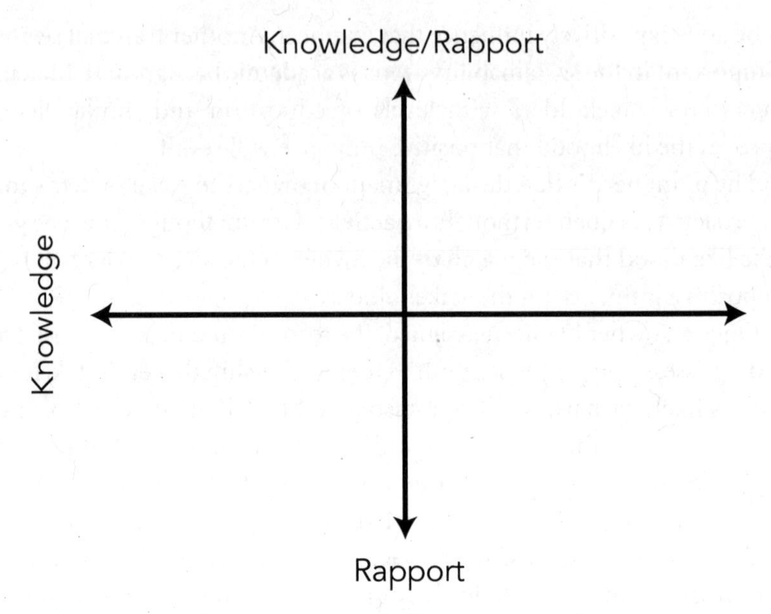

Figure 9-5 Knowledge over rapport.

Technical complexity may be fun and interesting to scientists and engineers, but it is the enemy when it gets in the way of helping a key stakeholder group understand your sustainability initiative.

The rapport dimension assesses the extent to which a stakeholder group feels positively or negatively toward your initiative. This is a measure of the emotion that your initiative evokes from a particular stakeholder group. The fact is that the topic of sustainability does evoke both positive and negative emotions in individuals and stakeholder groups. Working to assess the level of rapport that your stakeholder group has with your sustainability initiative and working to understand what drives that level of rapport are important to the team's ability to influence those stakeholders. Although it is more difficult to move an individual or group along on the rapport dimension, nonetheless, it is important to plan communications and carefully assign owners who may have the ability to have a positive impact on rapport.

An important principle, when building strategies to move stakeholders up in the knowledge and rapport dimensions, is that building knowledge is more important and impactful than trying to build rapport. Stakeholder groups will take positive action when they understand the rationale and benefits of your initiative, even if they have some negative emotional response to the initiative. For example, a simple action such as moving an organization to two-sided copying to reduce paper consumption creates some level of inconvenience for workers, but when the workers understand the positive impact to the environment and the costs versus benefits of the action, they are more likely to embrace the move.

The knowledge-over-rapport effect is especially true when dealing with decision makers. Teams can waste a great deal of time trying to get a CFO or a key regulator to like them. Realistically, the only way to influence a CFO is to present an accurate, positive business case. And the only way to influence a key regulator is to demonstrate compliance through accurate, timely data.

For these reasons, assessing where each stakeholder group is positioned in the dimensions of knowledge and rapport can help teams to build winning strategies and greatly affect the success of a sustainability initiative.

Likelihood/Degree of Stakeholder Support/Opposition to a Project and Associated Factors

The final factor that supports the development of stakeholder strategies and completion of the relationship-management plan is the extent to which stakeholders are likely to support or oppose a project or initiative and the factors that are working for or against that support or opposition. It is important for teams to make a realistic assessment of where the support for the sustainability initiative is and to identify the factors that drive that support. Similarly, they need to assess where the opposition is and identify the drivers of that opposition. This activity will help them to develop more creative strategies for building support and reducing opposition within each stakeholder group.

A practice that has been employed within the Six Sigma community can be helpful here. The practice is entitled *force-field analysis* (Figure 9-6). We can demonstrate and discuss force-field analysis by expanding on the activity we conducted earlier of analyzing the three basic stakeholder groups.

Force-field analysis is another activity that is driven primarily through team-based brainstorming and the prioritization of ideas. The activity is focused on asking two basic questions relative to each stakeholder group:

Stakeholder	Enablers *(What is working for us?)*	Restrainers *(What is working against us?)*
Customers	Our customer base has demonstrated that they will take positive action when faced with choices that can positively impact the environment.	Customer knowledge of our efforts in reducing our carbon footprint is very low and our advertising and communication budget is very limited.
Employees	Our generation X employees generally embrace team-based activities.	Senior management has been unwilling in the past to provide the necessary resources required for teams to be successful.
Shareholders	Other companies have demonstrated that successful sustainability initiatives can significantly reduce total costs and drive stock price values.	This initiative must compete for capital investment dollars versus other capital investment requirements. IT infrastructure investments seem to have priority over sustainability.

Figure 9-6 Force-field analysis.

- What is working for us?
- What is working against us?

Factors that fall into the "what is working for us" category are called *enablers.* Enablers are factors that are based on how the stakeholder group will benefit from the initiative or other reasons why the stakeholder group is likely to support the initiative. Enablers also can be factors that are already in place to help drive an initiative forward such as leadership support and approved investments.

Factors that fall into the "what is working against us" category are called *restrainers.* Restrainers are factors that are based on how the stakeholder group will be negatively affected by the initiative and other reasons why the stakeholder group is likely to oppose the initiative. Restrainers also can be factors or realities that are likely to become real barriers to the success of an initiative, such as a lack of leadership support or absence of investment capital.

By making a realistic assessment of the factors that will either enable or restrain a stakeholder group from providing support to the sustainability initiative, a team can develop informed strategies that will leverage the enablers and overcome or neutralize the restraining factors. The application of force-field analysis therefore becomes a useful element in developing an effective relationship management plan.

As a part of their collective experience and body of knowledge, Six Sigma practitioners have learned that key stakeholder groups play an important role in either the success or failure of any improvement initiative. For this reason, focus is placed on identifying who the stakeholders are likely to be and understanding what their expectations and requirements are. This reality is especially true with regard to sustainability initiatives. The web of stakeholder groups, ranging from self-appointed employee groups to government regulators, must be understood and managed carefully.

The purpose of this chapter is to provide readers with a framework and a set of practices for understanding the types of stakeholder groups and then developing proactive strategies for engaging and managing them. In the next section we will use this framework to analyze key stakeholder groups that are likely to influence your sustainability initiative.

The Web of Stakeholder Groups

In the field of environmental sustainability, there is no shortage of internal and external stakeholder groups. A high-level scan of both areas quickly reveals the complexities of issues and logistics in coordinating an overall influence strategy for achieving the company's sustainability-related business goals. Note, here, that we are not just focusing on technical improvements such as energy and water efficiency. After all, these are only in the circle of our environmental issues. Because we are interested in the overlap of both environmental and business-value-creation drivers, technical progress is insufficient (see Chapter 2). It is perfectly normal to have one set of stakeholders for achievement of the environmental value and a separate set of stakeholders for achievement of the business value—on the same issue. Consider the goal of improving waste-diversion rates. In order to make a sufficient case for a wide-reaching initiative in this area, Apex needed to tap into brand reputation as a business-value driver in addition to waste reduction as an environmental-value driver. Each part of this equation has separate stakeholders, as shown in Figure 9-7.

Issue	Environmental Stakeholders (Waste Diversion/Reduction)	Business Value Stakeholders (Brand Reputation)
Waste reduction	Employees	Public relations team
	Facilities management team	News media
	Municipal infrastructure managers	Various "watchdog" organizations
	Waste haulers	Green ratings programs

Figure 9-7 Issues-based stakeholder identification example.

Within the company, a mature sustainability program will affect every functional group. Your choices for resource allocation for outreach just will depend on your assessment of which group is most important at any given point in time. Some examples of internal stakeholder groups are provided in Figure 9-8 with comments about how they connect with environmental sustainability.

Most companies forming new sustainability programs will be in *retail mode,* which is to say that initiatives will be formed and stakeholder engagement accomplished more on a one-to-one basis. Your company even may already have a wholesale employee sustainability program before you charter a formal corporate program. These wholesale programs typically take the form of a green team's effort or a grassroots, volunteers-based program that may or may not be officially sanctioned by the company and may or may not have any influence over the environmental programs of the company. Green team programs can be a good, cost-effective investment for building general awareness of sustainability issues across the company. This sort of cultural awareness can create a tailwind that makes further change a bit easier. Once your program goes beyond general awareness, you also will need to shift from general engagement with employees as a group to engagement of employees based on their roles within the company. Improving sustainability performance within specific company functions requires more specialization on the issues inherent to the new behaviors you are trying to form. Although employee participation in green team programs eventually will plateau—and can run out of steam without some structured management and periodic events—they also can be a great way

Group	Sample Sustainability Decision Making Authority
Sales and Marketing	Ability to represent your company's products and services with a realistic sustainability message. Without a thoughtful process, the risk is this group will over-state the environmental benefits of your products (green-washing). Positions sustainability announcements and formal reports to the outside world; probably your own CSR reporting approvals.
Finance/Investor Relations	Incorporate the discounted financial value of sustainability impacts on investment decisions. Should understand what the company is doing to improve environmental performance and to mitigate the physical and regulatory risks to core business. Explain the shareholder impact of sustainability goals and progress to current and potential investors.
Procurement	Know which of your company's products use feedstock that is sensitive to environmental issues in the short and long term. Evaluate supplier businesses and their products for sustainability performance and consistency with your company's sustainability priorities. Represent supplier innovations and ideas for improving your company's foot print.
Product Teams	Use sustainability as a catalyst for internal innovation. Understand the environmental and social impact of the products they design. Understand trends in customer expectations for lowering environment impacts. Offer product substitutes or new products that are eco-friendly for existing or new markets.
Real Estate and Facilities	Manage the efficiency of the physical infrastructure of your company. Directly and indirectly reduce the energy and water use of facilities. Mitigate the climate change-related physical risks of facilities location decisions. Use green buildings for employee health drivers such as indoor air quality and for brand drivers related to certifications such as LEED. Work with local municipal resources to improve infrastructure.
Human Resources	Understand the role of sustainability as a drive of attraction and retention of talented employees in all employee demographics. Advise the company on skills development for building sustainability into corporate functions.
Information Technology	Manage the information and data infrastructure underlying business decisions that incorporate sustainability. Make purchase decisions for carbon management database technology. Make equipment and application purchases that reduce energy consumption. Locate datacenters in context of climate change weather risks for local geography.

Figure 9-8 Decision-making authority by stakeholder group.

to discover individuals within each corporate function that could help as subject-matter experts. For example, at Apex, the green team program recruited a manager from the Finance Department who was instrumental later in creating a tool that incorporated the price of carbon into company investment decisions. The tool became a great way to connect this person's passion for sustainability with his day job and create a sustainability champion within the Finance Department.

Outside your company, there are a number of stakeholder groups that represent the public good, consumer protection, and other issues-based constituencies. These organizations are formed to aggregate the influence of their members and focus that influence on actors who represent opportunities and threats to the organization's mission (Figure 9-9).

Outside organizations possess power over your company in proportion to their influence over your company's customers, suppliers, and employees. You might consider this power to be a function of the influence of the organization's resources (as a proxy for the strength of its "voice"), the breadth of the organization's issues base, and the geographic presence of the organization. In other words, if your company is a global technology company, then third-party organizations that have influence over your customers, work on a variety of sustainability technology issues, and are global would be worth your engagement at the corporate level. For example, the Global e-Sustainability Initiative (GeSI) was chartered in 2001 "to

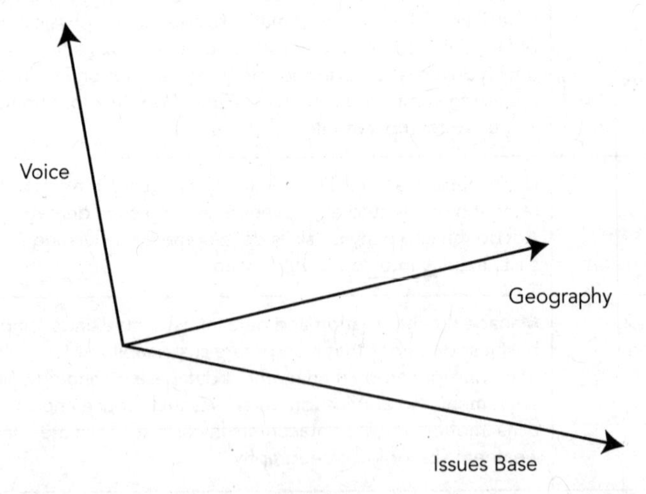

Figure 9-9 Three factors contributing to organization's influence.

further sustainable development in the ICT [information and communications technology] sector. GeSI fosters global and open cooperation, informs the public of its members' voluntary actions to improve their sustainability performance, and promotes technologies that foster sustainable development."[2] In this particular case, GeSI is an example of an association of companies from a single industry that serves as a focal point for information and ideas but not for the purpose of strongly influencing the purchase decisions of technology consumers. GeSI draws its strength by the association of leaders in the ICT industry and by its partnership with the World Wildlife Fund (WWF). Depending on the knowledge of its principles and the rapport struck with other influencers, it is likely to be of some help in championing initiatives with positive sustainability value and blocking initiatives that detract from this goal.

On the other hand, traditional conservation-oriented organizations such as the WWF are more focused on advocacy and advancing a social agenda. The history of the WWF and its status within the environmental field suggest a stronger level of influence over sustainability-minded customers around the world. Its corporate, government, and humanitarian partnerships are indicators of the value the WWF places on stakeholder engagement and of the role that credibility capital can play in sustainability.[3]

Advocacy organizations typically evolve based on an investment in expertise and the development of relationships to leverage that expertise to further a mission. The better job the organization does at these two functions, the more support it attracts, and the more its influence grows. By the time an advocacy organization achieves the status of a household name, it is likely to be focused on both its original targets for influence, typically government policymakers, and other sources of support, such as corporations. It influences the organizations to join its agenda, and this growing list of associated organizations increases its influence over policymakers, and so on. Organizations such as the WWF, the Nature Conservancy, and the Sierra Club have achieved this sort of status.

Just to show the breadth of issues addressed by organizations in sustainability, we list these examples of influential organizations and their associations:

- Ceres[4] represents financial investors to influence policymakers as well as public companies to achieve a low-carbon-economy future

- There are specialty organizations such as the Rocky Mountain Institute[5] (transition from coal and oil to efficiency and renewables), the Forest Stewardship Council[6] (forest management), and the U.S. Green Building Council[7] (green buildings)
- For water quality and conservation beyond global groups such as the WWF, there are organizations such as the Surfrider Foundation[8] (ocean protection), Salmon Safe[9] (land management to protect salmon in West Coast watersheds), and local riverkeeper groups (river protection).
- As collaborators for training and other labor issues, many companies work with community colleges and labor unions.

As stakeholder engagement programs are put into practice, your company will reach out to individuals from several organizations and will need to aggregate this feedback by sector. The aggregate information—often illustrated with direct quotes from individuals—is commonly shared in the reports that companies publish about sustainability efforts. As sustainability issues are increasingly honed as core business issues, information about progress on this front is integrated into corporate annual reports alongside financial data. As technology makes it easier to reveal performance data to the public, reporting performance data becomes more and more frequent, which feeds social media commentary as well as news coverage by specialty media such as GreenBiz.com and Triple Pundit.

Conclusion

Information collected from stakeholder groups, in addition to the expectations of customers about performance, becomes the input for the decisions made in the collaborative management model discussed earlier in this book (see Chapter 2). As the information is digested and performance is evaluated, representative messages should be extracted from the process for reporting and public discussion. We have found that more and more companies are forming persistent sustainability advisory committees to collect periodic advice from experts. Companies use these expert groups to discuss industry trends, ideas for new product innovation, or technical research needs. These groups might take one of three or more forms: general advisors to the CEO and chief sustainability officer, suppliers convened to address supply-chain issues, or employee focus groups. By tapping in to the knowledge of each set of experts, the company can integrate sustainability all along its value chain.

Chapter Summary—Key Points

- The number of stakeholders interested in the environmental performance of companies has increased dramatically over the past few years, as have the breadth of issues reviewed and the sophistication of the analysis.
- For a company to be considered an effective reporter of its sustainability progress, stakeholders expect transparency in both performance and process.
- Stakeholder management is the set of processes and practices that enables one to deeply understand, monitor, and influence the perceived needs of various groups or individuals who have the potential to slow or block your program. Stakeholders also can accelerate adoption of your sustainability program through positional or informal leadership.
- The relationship-management plan enables teams to understand who the key stakeholders are and how their needs will be addressed as the overall sustainability initiative moves forward. The strategy for influencing stakeholders is built on an understanding of their needs and also considers
  - The type and level of influence that the stakeholder has with regard to the sustainability project or initiative (e.g., decision maker, champion, or blocker)
  - The assignment of a key team member to own this relationship and to drive the strategy with this group or individual (balancing knowledge and rapport)
  - The strength of the relationship that the group or individual has with your organization or team
  - The extent to which the group or individual is likely to support or oppose the team's project or initiative and the factors that are working for or against that support or opposition (e.g., can be determined by applying tools such as force-field analysis)
- The global influence and knowledge of third-party sustainability stakeholder organizations such as the World Wildlife Fund, the Global

e-Sustainability Initiative, CERES, the U.S. Green Building Council, and the Surfrider Foundation are strong.

- Confrontation with stakeholder groups is an outmoded form of engagement. By working in collaboration with stakeholder groups such as employees, suppliers, and nongovernmental organizations, new, effective strategies can be designed and deployed around the world.

Notes

1. www.cisco.com/web/about/ac227/csr2010/governance-and-ethics/stakeholder-engagement.html.
2. www.gesi.org/?tabid=8.
3. www.worldwildlife.org/who/index.html. According to the group's Web site, the WWF's mission is the conservation of nature. The group's work includes protecting natural areas and wild populations of plants and animals, including endangered species; promoting sustainable approaches to the use of renewable natural resources; and promoting more efficient use of resources and energy and the maximum reduction of pollution.
4. www.ceres.org.
5. www.rmi.org/rmi/.
6. www.fscus.org/about_us/.
7. www.usgbc.org/.
8. www.surfrider.org/.
9. www.salmonsafe.org/about.

CONCLUSION

Letters to Tomorrow's Corporate Leaders

Letter to Tomorrow's Chief Sustainability Officer

Dear Ms. Curtiss:

Congratulations on your new position! The good news is that you have a seat at the table and that your company cares enough about sustainability to invest in a formal leadership structure to do something about it. The challenging news is that you are likely to be in a role with little to no direct control over the resources and operations of your company. You are also in charge of a function that is not very well understood or may be considered window dressing or both. We'd like to share with you some information that we think will help you with your new challenge.

Start by pledging to your executive colleagues that sustainability at your company will be a strategy for generating shareholder value. Ask them to please challenge any initiative that emerges from your program and to feel free to delay the initiative until together you can agree on the business case should you ever seem to stray from this philosophy.

Next, get going in formalizing your high-value program opportunities in energy efficiency. We have found that although many companies work on energy efficiency for years, a formal program with strong management still can achieve very good immediate results. You may need to encourage the development of new skills or tools within your facilities organization. But chances are that in your first conversations with corporate real estate leaders, you will find the basis for some very good success stories to share as well as some very good ideas that can be structured and taken to scale.

What are employees asking about? If your company is like others we know, employees probably wonder if your company is executing on the basics, such as recycling. You'll probably do an employee survey sooner or later and be pleasantly surprised that so many employees have so many great ideas about how to make the company more sustainable. The risk here is that you'll have hundreds of ideas from hundreds of employees that are now growing impatient wondering when their idea will be implemented. Think ahead about how you're going to filter ideas and keep employees in the loop.

Speaking of risk, this might be a good time to do an assessment of the risks your business faces relative to sustainability. Get your colleagues involved—the more the better! You want as many eyes on risk as you can get. Maybe you have some products that are eco-unfriendly. Maybe you have feedstock from your supply chain that is sensitive to flood (or to drought). And there may be regulations around the corner or around the world that are poised to make your company commit resources in unanticipated ways. Reactive is no way to run a business. In your Six Sigma handbook you'll find several tools for prioritizing risk—a failure modes and effects analysis can be formal and quantitative, or it can be a conceptual framework that gets you asking the right questions. After all, climate-change adaptation is just another word for contingency planning. And you'll find that climate-change effects on your business cry out for discipline in managing severity, occurrence, and detectability.

Now that you have some short-term wins on the horizon, start defining your company's sustainability transfer function. Work with employees, customers, and other stakeholders to prioritize issues based on the connection to your business. And assign financial costs to defects in each major area. Getting project teams assigned to the highest-priority, lowest-performing items will generate real progress and measurability. Get your data flowing. And prepare to share performance across the company and with the outside world.

Six Sigma and change management together can create amazing results. But this doesn't happen automatically. It takes your leadership. And if you want your company on board, it takes a connection with the core of your business. We think that Six Sigma provides a platform with the kind of rigor that you need for driving results to your triple bottom line. We are excited about the path you are on! And we're counting on you to inspire.

Sincerely,

Tom, Michael, and Dan

Letter to Tomorrow's Corporate Real Estate Leader

Dear Mr. Johnson:

As you know, a lot can happen when you have a seat at the company executive leadership table. Your influence over company direction grows based on your knowledge of labor markets and your ability to anticipate infrastructure needs. Your influence over company performance grows when you can show tens or hundreds of millions of dollars in cost reduction through portfolio rightsizing. No corporate real estate (CRE) executive gets there by being great at managing the efficiency of building systems. You get there by a relentless focus on improving business performance.

We're hoping that the concepts and perspective in this book adequately support the idea that sustainability is not only an important business issue but also one that can bolster your personal leadership profile across the company thanks to the specific role that you can play. Especially if you work in a services industry, no other senior executive in your company can have as much direct impact on your company's carbon emissions, waste reduction, and water footprint. This puts you within striking range of leading cultural transformation, risk reduction, shareholder satisfaction, and operating expense heroics.

Sustainability creates a reason to understand your company's customers better—yours is the organization best positioned to answer those surveys. It creates another reason to look at densities, location decisions, and capital management. Sustainability may not be the driving factor of your supply-chain transportation optimization project, but it creates another column for accumulating points when fuel usage and related carbon emissions are slashed.

And couldn't you use some employee buy-in for that higher-density open plan? For daytime cleaning? For your ride-share commute-reduction program? With you at the lead of an issue that grows in importance as the next generation of talent achieves higher levels of influence, you have the opportunity to inspire a new level of employee engagement, provide an environment that enables new levels of health and productivity, and add meaning to the company mantra that "our employees are our most important assets."

We've seen what started as internal-only green building technical innovations become products at some of the largest companies in the world.

CRE innovation can lead to top-line results, and sustainability is a platform for innovation.

We also know that there is a chasm you'll need to cross if you want your sustainability program to inspire. The chasm is between your initial focus on resource efficiency in facilities management (which is easy and under your direct control) and a new level of workplace productivity (which is more complex and cross-functional). The chasm can be crossed only by reaching out to colleagues—to the general employee population, to the information technology (IT) department, to sales and investor relations. The low-hanging fruit of sustainability is within your direct control. But the real payoff requires collaboration across business functions.

This takes knowledge, courage, and a willingness to enter into a discovery process that explores requirements from new customers.

And these two criteria—the need for discovery and the ability to affect the customer experience—are the two most important criteria for deploying the Six Sigma toolkit to a business challenge.

Corporate real estate executives are action-oriented. You are driven to achieve tangible results. This mind-set is a perfect trigger for leading once you make your own decisions about the business areas in which you want to invest and innovate.

We hope that sustainability is one of those areas.

Sincerely,

Tom, Michael, and Dan

APPENDIX A

Business Case Template and Examples

Below is an example template for creating an effective business case. The italicized text provides instructions for each section, and many sections include examples. It is important to note that the examples are condensed, and the actual business case should contain more detail than what is presented on this template.

Summary

Present the summary business case for the project. This section should articulate the overall objective, high-level process, and expected outcomes from the project. This should be no more than two to three paragraphs.

Example

Unchecked, Company A's energy expenses will rise to $1 billion over the next decade, and the overall CO_2 emissions will increase 60 percent to 8.0 million metric tons given a volume growth rate of 3 percent annually and an energy cost inflation factor of 4 percent.

It is being proposed that . . . [describe program].

Background

Present the background for why this business case is needed by the client. This section should focus on the background, provide industry best practices, and any initial findings. This should be no more than two to three paragraphs.

Example

Energy prices have risen drastically and become more volatile over the past decade. With added emphasis on sustainability, brand, and CO_2 emissions, energy efficiency is more important than ever to company operations. Efficient use of energy in facilities not only will help those facilities to be good stewards of the environment but also will assist them in lowering their cost of production and reducing their operating risk profile.

Statistics	Facilities	Other facilities	Entire portfolio
Current energy budget	$200,000	$600,000	$800,000
Percent of energy total	20.0%	80.0%	100.0%
Number of facilities	200	800	1,000

Over the past six months, a diverse mixture of X facilities has been assessed to gather data and validate the assumption included in the supporting analysis. The results from these assessments show the assumptions made in the attached analysis are at the very least conservative in nature (see table below for details).

Initial Facility Assessments

Number of facilities	Current energy spend	Annual avoidance	Investment required	Percent avoided	5-Year rate of return
6	$5,000	$1,789	$5,625	28.6%	21.4%
5	$9,000	$2,369	$8,000	24.5%	18.3%
1	$5,00	$1,250	$3,500	19.7%	26.8%
6	$8,000	$1,830	$5,445	22.9%	23.9%
2	$4,135	$823	$1,753	19.9%	41.3%
20	**$34,433**	**$8,060**	**$24,323**	**23.4%**	**23.3%**

Strategy

This section of the business case should present the overall strategy. This section should focus on the project's mission statement, key performance indicators, and an action plan for executing the project. This section should cover all important points of the project.

Example

Mission Statement

Team X is an internal consulting organization, centrally housed and managed within the Sustainability Department of the organization, which will ally the team with vendor partners to provide facility monitoring and assessment services. The team also will provide improvement-project development and management services to various owned and operated facilities and eventually the entire supply chain to improve the overall energy efficiency of the manufacturing process by 20 percent.

Tactically, the program will be deployed as follows:

1. First, an electronic energy-management system (EMS) will be installed at the candidate sites and monitored for a period of at least 60 days to evaluate the energy-usage profile of the facility. The facility workers then will be asked to fill out a detailed questionnaire and provide copies of the utility invoices and emissions permits.
2. Next, an on-site facility assessment will take place to interview the operating staff and evaluate the potential improvement measures to be implemented. The total rate of return for each project implemented will be targeted at a minimum of 17 percent over a 5-year analysis period (with exception of cogeneration, alternative-energy, and fuel-switching projects, which will be evaluated over a 10-year period to better account for the long-term nature of these types of specialized projects).
3. Then vendor partners will be deployed to develop firm proposals for the suggested improvement measures, and the projects will be implemented if the firm proposals agree with the initial analysis. These proposals will cover all project costs, including engineering and project-management expenses, along with contractual guarantees of long-term performance.
4. Finally, the EMS system will be used to measure and verify the performance of the installed improvement measures over time to ensure that the desired results are achieved.

As soon as the program is fully developed and the concept proven at operated facilities, it can be expanded quickly to the remainder of the portfolio. The program also can be expanded easily to include the development of additional productivity initiatives discovered during the assessments process, such as water-use reductions and production efficiency enhancements, with the overall goal continuing to be a total IRR of 17 percent.

Capital Investment

The capital investment section should detail all costs and savings associated with the business case based on the overall strategy mentioned in the preceding section. The focus of this section is on the project's capital investment requirements and avoidance/savings potential. This section should cover all important points of the project related to finance.

Example

Based on an estimated average facility energy efficiency improvement of 20 percent in relation to the current company-wide energy spend for manufacturing of $XXX (of which approximately 20 percent is for facilities), the capital spend shown below will be required to implement over the next five years. The capital budget includes funding for EMS system deployment and improvement-measure implementation. If the program is expanded to noncompany partners, those facilities either can provide the capital through their own budgets or funds will be made available through loans and paid back by the facilities through the cost avoidance the implemented projects generate.

10-Year Capital Investment Summary

000s $US	Facilities	Nonfacilities	Entire portfolio
Program development	$0	$0	$0
Energy management systems	$20,000	$85,000	$100,600
Project investments	$80,000	$400,000	$400,131
Total	**$100,000**	**$485,000**	**$500,731**

Operating Costs

This section should detail all impacts on operating costs for the entire project.

Example

The project will involve numerous vendor partnerships, the creation of a program manager position, and two energy-efficiency manager positions to help manage the program.

Operations Budget Request by Year

000s $US	2007	2008	2009	2010	2011	Totals
Program development	$100	$0	$0	$0	$0	$100
Salary and benefits	$300	$600	$600	$600	$600	$2,700
General expenses	$10	$15	$15	$15	$15	$70
Travel	$100	$300	$300	$300	$300	$1,300
Consultant budget	$300	$1,200	$1,500	$1,500	$100	$4,600
Currently budgeted						
Total	**$810**	**$2,115**	**$2,415**	**$2,415**	**$1,015**	**$8,770**

Risks

This section should outline overall project risks and the mitigation plan.

Example

Failure to Deliver Energy-Efficiency Results

Only projects proven to be effective will be implemented. Beta projects to prove the viability of individual technologies and initiatives will be performed as needed before they become part of the standard portfolio of improvement measures investigated during the facility assessments. Additionally, vendor partners will be required to guarantee the performance of their projects.

Sensitivity

Within the business case, a sensitivity analysis should be conducted to test the validity of the baseline assumptions. This section will focus on any of the areas where the team feels sensitivity analysis would be appropriate.

Example

A general inflation rate of 3 percent, along with an energy inflation rate of 4 percent, was used in the analysis. Additionally, no value for the carbon credits generated was considered given the less than fully developed global market for carbon credits. The tables below show the resulting change to the internal rate of return (IRR) given changes in the assumed inflation rates and carbon-credit pricing.

Inflation Rate Sensitivity Analysis

IRR, %	**Annual Volume Increase**		
Energy inflation rate	**2.0%**	**4.0%**	**6.0%**
Flat	11.2%	14.3%	17.5%
2.5%	15.4%	18.6%	21.9%
5.0%	19.6%	**23.0%**	26.3%

Base case: 3 percent volume increase and 5 percent energy inflation without any carbon-credit value

Carbon-Credit Value Sensitivity Analysis

IRR, %	**Current year carbon-credit value**		
Carbon-credit inflation rate	**$5.00**	**$10.00**	**$20.00**
Flat	23.9%	24.8%	26.5%
2.5%	24.0%	25.0%	27.1%
5.0%	24.2%	25.4%	27.7%

Base case: 4 percent volume increase and 5 percent energy inflation

Deployment Schedule

This section should present a high-level project schedule with key milestone dates.

Example

August 07 Program approved and funded.
Begin final program development and roll-out at initial sites with chosen vendor partners.

	Budget for 2008 capital appropriation.
August 08	Update management team on progress and request appropriation of following year's capital. Begin to offer program to other partners.
Moving forward	Program is maintained and modified as needed to remain effective.

Total Value of Funding Request

Display a detailed summary of the funds being requested, including both operations and capital dollars.

Example

Funding Request Summary by Year

000s $US	2007	2008	2009	2010	2011	Totals
Capital budget	$6,000	$20,000	$28,000	$35,000	$15,000	$104,000
Operating budget	$1,000	$2,000	$2,000	$2,000	$1,000	$8,000
Total	**$7,000**	**$22,000**	**$30,000**	**$37,000**	**$16,000**	**$112,000**

Accountabilities

This section should list all executive sponsors, stakeholders, and key project team members with their roles/responsibilities during the project.

APPENDIX B

Sustainability Transfer Function

As covered in Chapter 3, the *transfer function* is a valuable Six Sigma tool because it shows the chain of cause and effect for a particular result. The transfer function can be derived quantitatively, as when it is calculated by a regression analysis. It can also be extrapolated from research and logic, as is the transfer function on the opposite page. This graphic shows a decomposition of "Sustainability" into its component parts, then breaks down those components, and so on. There are important semantic rules when labeling the parts of a transfer function:

- ▲ The "Big Y" (Y) is the label of the primary result being described in the transfer function. This is a result that cannot be directly manipulated, e.g., stock price or carbon footprint.
- ▲ The "Little y's" (y's) of a transfer function are subordinate results. These items are also not directly controllable. They are listed as y's because they are subordinate to a Y. E.g., revenue is a y but is subordinate to the Y of stock price; or, in our example, Physical Degradation of the Environment (y3) is subordinate to Environmental Sustainability (Y2) which is in turn subordinate to sustainability (Y).
- ▲ The "Big X's" (X's) of a transfer function are generally categories of outputs such as geography or demographics. They are descriptive for comparison and might be controllable but they are not considered causes that one would manipulate to influence performance.
- ▲ The "Little x's" (x's) are what we are after in a transfer function. When x's such as procurement practices or specific inputs are identified, then we know we have line of sight to the means of improvement.

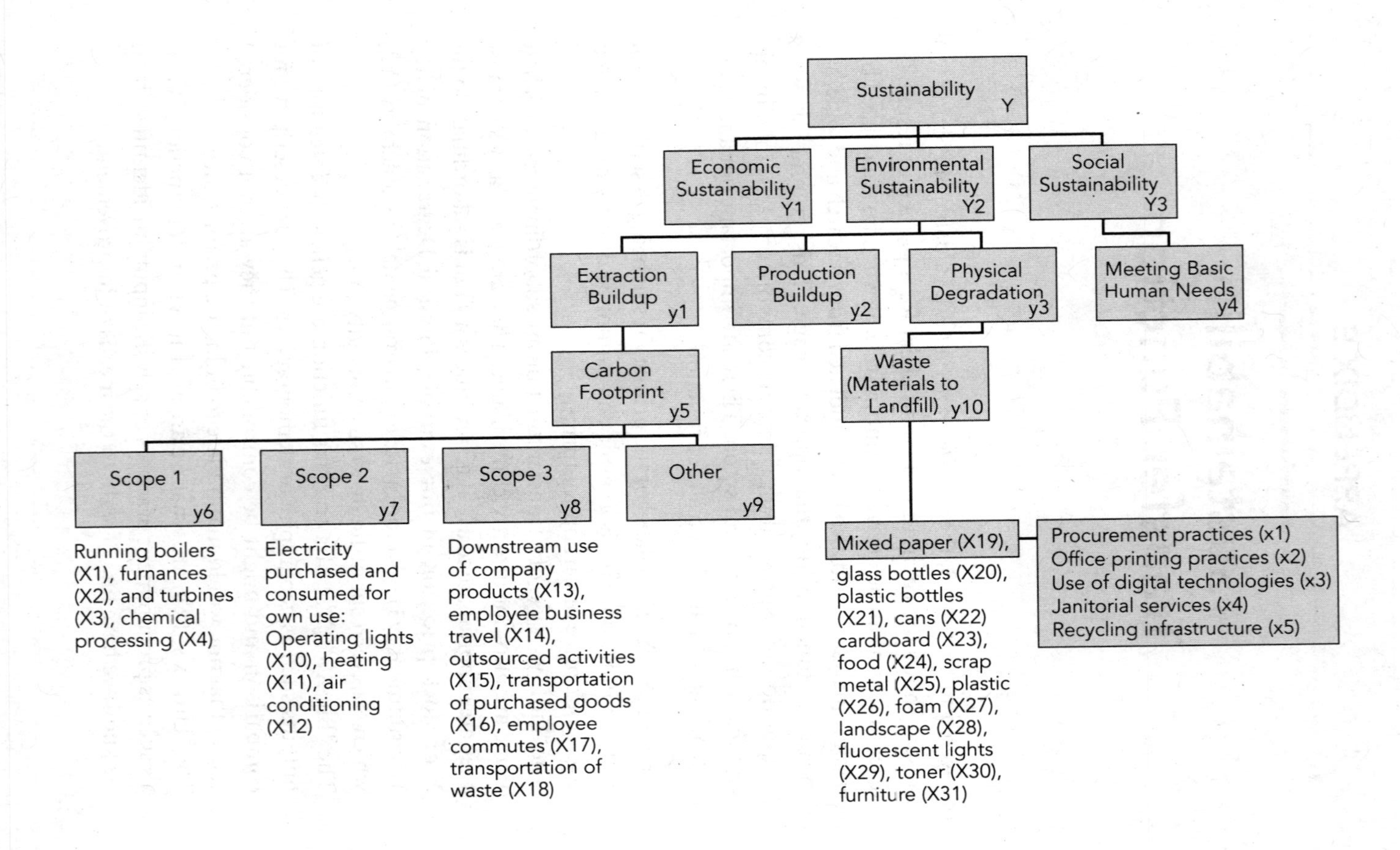
Sustainability Y
Economic Sustainability Y1
Environmental Sustainability Y2
Social Sustainability Y3
Extraction Buildup y1
Production Buildup y2
Physical Degradation y3
Meeting Basic Human Needs y4
Carbon Footprint y5
Waste (Materials to Landfill) y10
Scope 1 y6
Scope 2 y7
Scope 3 y8
Other y9
Running boilers (X1), furnances (X2), and turbines (X3), chemical processing (X4)
Electricity purchased and consumed for own use: Operating lights (X10), heating (X11), air conditioning (X12)
Downstream use of company products (X13), employee business travel (X14), outsourced activities (X15), transportation of purchased goods (X16), employee commutes (X17), transportation of waste (X18)
Mixed paper (X19),
glass bottles (X20), plastic bottles (X21), cans (X22) cardboard (X23), food (X24), scrap metal (X25), plastic (X26), foam (X27), landscape (X28), fluorescent lights (X29), toner (X30), furniture (X31)
Procurement practices (x1)
Office printing practices (x2)
Use of digital technologies (x3)
Janitorial services (x4)
Recycling infrastructure (x5)

APPENDIX C

Sample Energy Conservation Opportunity Evaluation Checklist for an Office Building Assessment

- Operations and maintenance:
  - Ensure that all equipment is functioning as designed.
  - Calibrate thermostats.
  - Adjust dampers.
  - Implement janitorial best practices.
- Occupants' behaviors:
  - Turn off equipment.
  - Institute and energy-awareness program.
  - Purchase ENERGY STAR equipment.
  - Install power-management software.
  - Institute policy to harvest daylight.
  - Install workstation task lighting.
- Lighting:
  - Change incandescent lights to compact fluorescents (CFLs).
  - Convert T12 to T8 and T5.
  - Delamp.
  - Install full-floor lighting sweeps.
  - Install occupancy sensors.
  - Install and use bilevel local switches.
  - Install high-efficiency light-emitting diode (LED) exit signs.
  - Install timer controls.

- Controls:
  - Adjust temperatures for comfort.
  - Evaluate and combine after-hours usage.
  - Adjust ventilation.
  - Limit access to thermostats.
  - Optimize startup times.
  - Adjust thermostats for seasonal changes.
  - Coast last hour of operations.
- Equipment:
  - Install heat-recovery equipment.
  - Relocate thermostats.

APPENDIX D

Sample High-Level Process Map for Energy Conservation in an Office Facility

A SIPOC is a useful tool for documenting a process at a high level, which is a common deliverable early in a DMAIC project. The SIPOC gets its name from its structure, which identifies the suppliers, inputs, process steps, output, and customers for a process. This tool helps to define scope for further analysis and, when its development is facilitated in a group setting, helps to align a team against a workflow—sometimes for the first time.

When we define a SIPOC process map in a team workshop, we follow these simple but important rules:

1. Start with identifying the customer—try to identify a single customer as the group that pays for the output of the process. Since customer requirements are foundational knowledge for Six Sigma, this is a good first step (and sometimes is a topic of great debate among a team!)
2. Identify the output that the customer cares about. In our example, we have a conceptual output of a "verified energy change." This is a bundle of action that creates energy conservation.
3. Identify the 5 to 7 (no more) high level steps that are the basic process flow. This is sometimes best facilitated by first identifying the start and end points of the process, then identifying the actions in between.
4. Now, for each process step, identify the inputs for the step and name the supplier of each input. A common mistake at this point is to make a generic list of inputs and suppliers and not mapping the former to the latter. If you don't know where the inputs come from, you don't know whom to ask to improve!

Suppliers	Inputs	Process	Output	Customers
Consumption tracker database (monthly) Facility management "Energy Star" ratings	Consumption data Expected consumption rates Benchmark data	**1. Determine gap in energy performance**		
Energy program lead: industry practices (EPA, BOMA, PG&E, etc.) Vendors Auditors or vendors Vendors, cleantech trend news	Standard energy O&M guidelines Gap assessment Results of energy assessments Information about new technologies	**2. Develop conservation idea**		
Facilities management Energy program lead with FM Local FM lead Facilities engineering Facilities governance	Facility suitability Cost of solution Expected energy savings Local energy rates Local power "dirtiness" Technology maturity Reputational impact	**3. Assess feasibility of idea**	Verified energy change	Senior executives Facilities management Facilities engineers
Facilities engineering: FM Vendors Facilities management	Basic implementation guides Engineering review Facility features	**4. Complete technical design**		
Facilities management Facilities engineering, vendors Normal budget process Facilities management	Facility information Technical expertise/labor Project management Funding Communications	**5. Deploy conservation measure**		
Vendor who did work Facilities engineering Energy program lead, engineering	Technical review Approved/proper verification protocols	**6. Verify conservation results**		

APPENDIX E

Sample Functional Performance Criteria for Enterprise Carbon Accounting Software

1. *Data display.* Is there a reasonable interface for displaying the data in the database? Can the charts and graphs be customized based on timeline, geography, facility, production process, or customer?
2. *Scenario planning support.* Can the software answer the important "what if" questions for your business? For example, what if your business expands or if you sell an operating unit? What if the price of energy rises in certain parts of the world? What if your company invests in on-site renewable power in certain locations?
3. *Data feeds.* Can the software accept data from other sources such as utility companies or smart meters? Can the software pass data to reporting frameworks such as the U.S. Environmental Protection Agency's (EPA's) ENERGY STAR Portfolio Manager?
4. *Predetermined reporting.* Can the software organize a report against common frameworks such as the U.K. Carbon Reduction Commitment, the Carbon Disclosure Project, or the Australian National Greenhouse and Energy Reporting System (NGERS)?
5. *Workflow management.* Can the software store information about improvement projects or other decisions that need to be made, manage a system of approvals for these projects, and show the status of each project in a pipeline of improvements?

6. *Measures and conversions.* Does the software support multiple currencies and multiple units of measurement for energy, water, and waste? Are carbon-emissions factors automatically applied for the energy source referenced? Can these emissions factors be manually overridden if need be?

INDEX

References to figures are in italics.